ROUTLEDGE LIBRARY EDITIONS:
ART AND CULTURE IN THE
NINETEENTH CENTURY

Volume 2

ENGLISH OPERA FROM 1834 TO 1864 WITH PARTICULAR REFERENCE TO THE WORKS OF MICHAEL BALFE

ENGLISH OPERA FROM 1834 TO 1864 WITH PARTICULAR REFERENCE TO THE WORKS OF MICHAEL BALFE

GEORGE BIDDLECOMBE

Routledge
Taylor & Francis Group
LONDON AND NEW YORK

First published in 1994 by Garland Publishing, Inc.

This edition first published in 2019
by Routledge
2 Park Square, Milton Park, Abingdon, Oxon OX14 4RN

and by Routledge
711 Third Avenue, New York, NY 10017

Routledge is an imprint of the Taylor & Francis Group, an informa business

British Library Cataloguing in Publication Data
A catalogue record for this book is available from the British Library

ISBN: 978-1-138-35894-2 (Set)
ISBN: 978-0-429-42671-1 (Set) (ebk)
ISBN: 978-1-138-36494-3 (Volume 2) (hbk)
ISBN: 978-0-429-43099-2 (Volume 2) (ebk)

ENGLISH OPERA FROM 1834 TO 1864 WITH PARTICULAR REFERENCE TO THE WORKS OF MICHAEL BALFE

GEORGE BIDDLECOMBE

GARLAND PUBLISHING, INC.
NEW YORK & LONDON / 1994

Library of Congress Cataloging-in-Publication Data

Biddlecombe, George, 1945–
English opera from 1834 to 1864 with particular reference to the works of Michael Balfe / George Biddlecombe.
p. cm. — (Outstanding dissertations in music from British universities)
Revision of the author's thesis (Ph. D.)—University of London, 1990.
Includes bibliographical references and index.
ISBN 0-8153-1436-1
1. Opera—England—19th century. 2. Balfe, M. W. (Michael William), 1808–1870. Operas. I. Title. II. Series.
ML1731.4.B5 1994
782.1'0942'09034—dc20 94-26321
CIP
MN

Printed on acid-free, 250-year-life paper
Manufactured in the United States of America

Contents

Preface to the Garland Edition

This is a revised version of my PhD thesis (University of London, 1990).

The thesis was in two volumes. The material has now been reformatted according to the requirements of the Garland series, including a new preface, acknowledgements, some alterations to the music examples, and an index. I have omitted a brief abstract and an appendix that dealt with Cecil Forsyth's *Music and Nationalism* (London, 1911).

The text is largely unaltered. I have, however, taken the opportunity of expanding certain points, such as Balfe's word-setting in 'I dreamt that I dwelt in marble halls' (chapter 7), details of the manuscript of Wallace's *Lurline* (chapter 8), and the impact of Wagnerism on English opera (Conclusion). Likewise I have expanded the information given in various endnotes. I have also corrected a number of typographical errors.

George Biddlecombe

Royal Academy of Music
February 1994

Acknowledgements

I wish to repeat the thanks expressed in the preface of the earlier version of this work to my supervisors, Mr Julian Budden and Dr Simon McVeigh.

Examples 2.1, 4.1, and 7.16.b are reproduced with the permission of The British Library.

I am grateful to the Royal Academy of Music for the use of various facilities and other assistance in preparing this revision.

Introduction

This thesis sets out to investigate English opera from 1834 to 1864. It is, therefore, unusual. By comparison with that of other countries English nineteenth-century music has received minimal attention. There is an assumption that, if not entirely 'das Land ohne Musik,' England contributed little music of value.[1] This view has existed for well over a hundred years.

To trace the development of this contention would require a separate study. Regarding opera, among the first commentators to call for a reassessment was Eric Walter White.. White's most detailed work concerned Britten and Tippett, but he drew broader attention to the subject in *The Rise of English Opera* (London, 1951) and *A History of English Opera* (London, 1983). These are general historical accounts rather than detailed studies. The *Register of First Performances of English Operas and Semi-Operas* (London, 1983) that White compiled in conjunction with his later study has been very useful for the present research.

A particular place in the study of English nineteenth-century opera must be accorded to Nicholas Temperley. His contribution has included mounting performances of Loder's *Raymond and Agnes* in Cambridge in 1966. This led to two articles, 'Raymond and Agnes,' *Musical Times*, 107 (1966), 307-10, and 'The English Romantic Opera,' *Victorian Studies*, 9 (1966), 293-301. Subsequently he has edited two volumes, *The Romantic Age 1800-1914*, volume 5 of *The Blackwell History of Music in Britain*, 6 volumes projected (Oxford, 1988),[2] and *The Lost Chord: Essays on Victorian Music* (Bloomington and Indianapolis, 1989), which include the most scholarly studies of the subject to date: in the former volume, Bruce Carr, 'Theatre Music: 1800-1834' (288-306); Michael Hurd, 'Opera: 1835-1865' (307-29); Nigel Burton, 'Opera: 1865-1914' (330-57); in the latter, Nicholas Temperley, 'Musical Nationalism in English Romantic Opera' (143-57).

Arthur Jacobs's monograph on Sullivan is a rare case, concentrating on a single composer and illuminating the musical scene of his time.[3] Regarding the type of work on related issues that is now increasingly common in the study of eighteenth-century English music, Leanne Langley's research into aspects of musical journalism is still isolated.[4]

Despite the research noted above, English opera of the period 1834-1864 is still little known. Yet these three decades saw an extraordinary expansion in the genre, initiated by Barnett's *The Mountain Sylph* of 1834. Balfe's first English opera dates from 1835. The year 1864 saw the last of his theatrical works to be given its premiere during his lifetime; the last of Benedict's; and the last of Macfarren's. It was the end of an era of considerable creativity. The purpose of this thesis is to bring this phase of operatic history to attention.

A particular feature of the present study is its broad contextual approach. This is necessary in order to understand the circumstances influencing the development of English nineteenth-century opera, its characteristic features, and the reasons why these traits held sway. The approach involves an investigation into the theatrical situation, the cultural factors shaping the libretto, and the techniques used by librettists, before embarking on a study of the musical elements. The latter must also be seen in context: we need to be aware of the national heritage (a very slender one) that fell to composers of English opera in the mid 1830s and of contemporary foreign opera, so that we may understand why they took certain decisions regarding musical styles and assess their contribution in this light.

The following account is therefore divided into two main areas. Chapter 1 concentrates on the theatrical conditions, such as the theatres available for English opera, legislation, and managements, together with the main personalities involved. Chapter 2 is concerned with the conventions of English libretti, primarily the use of melodramatic plots and characterization based on the passions. The ballad, a characteristic feature of the genre, is given special consideration here. The craft of writing libretti--the choice of subject and the adaptation of foreign works, particularly French--is discussed in chapter 3. Chapter 4 provides an account of the general operatic background. This deals with the English attitude towards opera, the salient features of indigenous operatic works prior to 1834, and a survey of the composers active from 1834 onwards.

The second part of the thesis investigates the work of the six main composers of the period--Balfe, Barnett, Benedict, Loder, Macfarren, and Wallace. Balfe's operas are given most consideration, for he was the most prolific. The focus is on operatic styles. With the exception of Macfarren's mature work, the native element was rarely represented by more than the ballad and the glee. Italian and German influences, however, were strong: there are clear echoes of Rossini, Bellini, Donizetti, and Weber. To a lesser extent there are signs of the French style. All need to be identified in order to put the musical aspect of English opera into sharper relief. Each composer is studied in the same fashion: details of career and output are followed by a study of the style adopted, focusing on works selected for this purpose. The discussion is concluded by a summary. Some account has been taken of post-premiere revisions, though this cannot claim to be exhaustive.

The conclusion summarizes the above findings. It also considers the prevailing view of English nineteenth-century opera, placing this alongside contemporary reports and offering an explanation for the change in opinion that has occurred during the last one hundred and thirty years regarding these works, some of which were among the most popular English operas ever produced.

NOTES

1. The term derives from Oscar Schmitz, *Das Land ohne Musik. Englische Gesellschaftsprobleme* (Munich, 1914) - a study, we may note, of Britain, but not its music.

2. Originally available as volume 5 of *The Athlone History of Music in Britain* (London, 1981).

3. Arthur Jacobs, *Arthur Sullivan: A Victorian Musician* (2nd ed., Aldershot, 1992).

4. Leanne Langley, 'The English Musical Journal in the Early Nineteenth Century' (unpublished Ph.D. dissertation, University of North Carolina at Chapel Hill, 1983); 'The Life and Death of The *Harmonicon*: an Analysis,' *R[oyal] M[usical] A[ssociation] Research Chronicle*, 22 (1989): 137-63.

English Opera From 1834 to 1864 with Particular Reference to the Works of Michael Balfe

I

The Theatrical and Literary Background

With rare exceptions, English opera has always been focused in London. The United Kingdom has never boasted the number of opera houses found in Italy or Germany. In the nineteenth century there was a closer parallel with the French concentration on Paris, but there were fundamental differences due to the repertories of the major opera houses in the two capitals and the State's financial support in France.

London had no theatre specifically designated for English opera on which native composers could rely. Foreign operas, traditionally given in Italian, were performed at the King's Theatre in the Haymarket. The theatre of 1789, successor to the original building of 1705, was renamed Her Majesty's on Victoria's accession in 1837. Patronized by the aristocracy, it was often known simply as 'the Italian Opera,' and had long attracted musicians of international stature. In the period under discussion, its season, lasting four to five months, was timed to coincide with the aristocratic 'season' and with the availability of outstanding singers. After a short initial period in March, employing less prestigious names, it started in earnest after Lent, when the stars of the day were released from their Paris engagements.[1] Luminaries such as Grisi, Mario, and Lablache sang regularly. The most famous composers to appear at the theatre were Rossini, in 1824, and Verdi, who directed the premiere of *I Masnadieri* (1847). By this time the theatre's fortunes had been severely weakened by internal dissent, provoking the conductor Michael Costa and a number of singers to form the 'Royal Italian Opera' at Covent Garden. Balfe took Costa's place, conducting the two performances of *I Masnadieri* that followed Verdi's departure. Her Majesty's was destroyed by fire in 1867, and a third building was demolished in 1892.

The main venues for opera in English, whether original works or translations, were Drury Lane and Covent Garden. In the nineteenth century their seasons, though variable, usually lasted from early autumn to the following summer--from September to July, for instance. Drury Lane owed its existence to a royal charter of 1663; Covent Garden first opened in 1732. Each had long offered a wide range of entertainment, including spoken theatre as well as musical works. Indeed, the two theatres were considered the guardians of the national Shakespearean legacy, and until 1843 enjoyed the increasingly dubious privilege of a legal monopoly on spoken drama, distinguishing them from the host of minor theatres whose numbers grew as London's population expanded apace. Both were casualties of fires in the nineteenth century--Covent Garden in 1808 and Drury Lane in 1809. The rebuilt Covent Garden was again destroyed by fire in 1856, and the present building dates from 1858. Drury Lane has remained largely unaltered since 1810.

The audiences at both theatres were largely middle-class, with poorer aficionados in the galleries, many of whom entered after the beginning of the evening's performance, when admission prices were reduced. The general demeanour of the audience reflected the tone of society at large. In the 1820s, many potential supporters of the theatres were discouraged from attending by the frequent bad behaviour, particularly from the galleries. Pückler-Muskau, a German prince who visited London at this time, recorded the social division between audiences at the Italian Opera (and the rewards for its foreign singers--'all flocking to English gold') and the two patent theatres. He was shocked by the low standards of conduct, and pointed to a specific deterrent against larger middle- and upper-class support:

> Another cause for the absence of respectable families is the resort of hundreds of those unhappy women with whom London swarms. They are to be seen of every degree, from the lady who spends a splendid income, and has her own box, to the wretched beings who wander houseless in the streets. Between the acts they fill the large and handsome 'foyers', and exhibit their boundless effrontery in the most revolting manner.
>
> It is strange that in no country on earth is this afflicting and humiliating spectacle so openly exhibited as in the religious and decorous England ... these are the scenes, I repeat, which are exhibited in the national theatre of England.

Sir Walter Scott confirmed this: 'No man of delicacy would wish the female part of his family to be exposed to such scenes; no man of sense would wish to put youth, of the male sex, in the way of such temptation.'[2] For the very religious--many non-conformists, for example--such reports simply emphasized the institutional immorality of the theatre. Distrust of the theatre's morals continued in some quarters throughout the century.

By the early 1830s there was a strong reaction against rowdy behaviour. The profligate ways of the Regency and George IV's reign were swept aside by the influence of William IV's more sober outlook. Within two years of William's coronation in 1830, the translator of Pückler-Muskau's recollections could note that

> since the foregoing Letters were written, the character of the higher society of England has undergone a considerable modification. The frank, noble, and practical turn of mind of the present King, and the simplicity, amiableness, and sterling excellence of the Queen, have broken the fool's sceptre of Fashion, and have introduced a better and higher standard of merit and of elegance than was recognized in the last reign.[3]

The royal example permeated not only 'higher society' but the middle and other classes. Prostitution came under control in the late 1830s and early 1840s, when the actor Macready managed Covent Garden and Drury Lane. The pater familiae could take his flock to the theatre without embarrassment. As the model of Queen Victoria and Prince Albert established itself, so audiences settled further into staid habits. The advertisements that start to appear in libretti in the 1860s were aimed at a range of middle-class families, from those whose incomes might direct them towards ready-made clothes rather than the more expensive hand-made dresses and suits, to others who could enjoy more expensive luxuries, such as imported wines, and the services of eminent doctors.

These social improvements were not matched by any fundamental changes in the circumstances of English composers. They were faced by a constant impediment: the lack of a secure theatrical base. In 1833 the composer George Rodwell (1800-1852) proposed an imaginative scheme in 'A Letter to the Musicians of Great Britain; Containing a Prospectus of Proposed Plans for the Better Encouragement of Native Musical Talent, and for the Erection and Management of a Grand

National Opera in London.' A pupil of Sir Henry Bishop and Vincent Novello, Rodwell had first-hand experience of the situation: by 1833 he had written music for thirteen theatrical works, and he tellingly contrasted the conditions under which British composers worked with those enjoyed by their continental counterparts. His plans were doomed by their projected costs.

Money was at the heart of the matter. None of the theatres received government subsidy--a constant source of lament. In the words of Benjamin Lumley, who managed Her Majesty's from 1842 to 1858, 'An operatic manager in England is consequently obliged to take the whole risks of a vast speculation upon his own shoulders, and is subjected to certain difficulties unknown in foreign countries.'[4] In 1840, shortly after being bankrupted as manager of Drury Lane, Alfred Bunn contrasted the situation in Paris, where

> the *Académie Royale*, the *Théâtre Français*, the *Opéra Italien*, and the *Opéra Comique*, have each an allowance from the government. The first receives no less a sum, per annum, than 800,000 francs (32,000*l*.) and the gratuitous use of the theatre besides--an amount, taken together, equal to the ordinary receipts for an entire season at either of our principal theatres; and after a certain period of service, its leading performers have a handsome retiring pension from the state, and their latter days are not subject to the limited means of a Theatrical Fund.[5]

Added to this was a morass of legalities concerning leases, patents, and shares of ownership, that further undermined the precarious financial stability of successive managements.[6] Vast fortunes were lost. When Captain Polhill, who had taken the lease of Drury Lane in 1831, passed it to Alfred Bunn in 1833, he was £50,000 poorer. By the time of his bankruptcy in 1840, Bunn had lost at least £23,000. In 1841 Macready took the lease, a decision that cost him £8,000 in one year; he left in 1843.[7] Bunn then returned, surviving until 1847. He was followed by the flamboyant conductor and impresario, Jullien, who engaged Berlioz as conductor for six years. His company lasted three months. Berlioz described Jullien as an 'undisputed lunatic' who had no idea how to run a company.[8] Not until the mid 1850s was the theatre's survival assured. Managements, including Bunn's and Macready's, came and went even quicker at Covent Garden. The repertory varied according to the management. Macready concentrated on spoken theatre, paying scant attention to his opera company. For much of the 1840s the building was silent, until Costa and the other secessionists from Her Majesty's took the theatre in 1847. From 1858 to 1864 Louisa Pyne and William Harrison leased the house each winter for their 'Royal English Opera Company.' Having begun their partnership in 1857, using the Lyceum, they moved when the new building became available, following the reconstruction after the fire of 1856.

The managements that commissioned most English operas were Arnold's, Bunn's, and Pyne and Harrison's. S.J.Arnold, son of the composer Samuel Arnold, reopened his 'New Theatre Royal, Lyceum, and English Opera House' in 1834. Between July 1834 and September 1835 Loder's *Nourjahad*, Barnett's *The Mountain Sylph*, and works by Lee, Macfarren, Packer, Rodwell, and Thomson were performed. But after two seasons he went bankrupt, having lost £4,000.

Three musicians who hoped for managerial success--the tenor John Braham and the composers Barnett and Balfe--also came to grief in the early part of the period, having achieved less than Arnold. Braham undertook the management of the St James's from 1835 to 1838; his hopes of performing *Oberon* (he had been a member of the original cast) foundered on questions of copyright involving the playwright and librettist, James Robinson Planché.[9] Barnett managed the same theatre, temporarily renamed The Prince's, for a brief time in 1840. Balfe's chastening experience came at the Lyceum in 1841. His enterprise was planned to include a group of composers, particularly Barnett, Loder, and Macfarren, but it collapsed, due to internal dissent, within a few months (March--May). In 1849 Balfe joined a committee that called for a national opera, but predictably this came to nought. John Maddox was more successful, encouraging English composers at the Princess's from 1843 to 1850.

The extraordinary Alfred Bunn deserves special attention. He had three periods of administration at Drury Lane: 1833-39, 1843-47, and 1851-52. In Planché's words, he was 'a strange compound'; a pugnacious, coarse man, who could pose as the young Napoleon on the Drury Lane stage during performances of Auber's *Gustavus III*. He made no secret of his 'utter detestation of Mendelssohn's music.'[10] Bunn acquired more notoriety than any other individual in the theatrical world during the period. However, he was driven by a burning desire to establish as close an equivalent to the Paris Opéra as possible.

Bunn's role in the development of English opera was very important. He supported two attempts by Balfe to institute all-sung opera: *Catherine Grey* and *The Daughter of St Mark*, writing the latter's libretto himself. *The Times* remarked on *Catherine Grey*'s 'daring and novel character' and explained that the opera 'is constructed altogether on the foreign principle, the whole of the dialogue being in recitative. An attempt of this kind, as it was in defiance of English prejudices, is extremely hardy.'[11] His fortunes were at their height in the mid 1840s. Musicians such as the soprano Adelaide Kemble and her father rejoiced 'at your success in the establishment of a real English Opera.'[12] By the time of the first performance of Lavenu's *Loretta* in 1846, *The Times* could wax lyrical on the 'progressive state of our national musical drama' to a wildly overenthusiastic degree: 'As the lyrical drama of Italy declines, that of England seems to rise, and the ultimate extinction of the one appears likely to be the epoch of the other's arriving at perfection.'[13]

All this was at the cost of intense dislike by the literary fraternity. There were vociferous complaints that Bunn subverted a bastion of the Shakespearian heritage, demeaning it with meretricious animal shows--criticisms that were not diminished by Queen Victoria's visits to see lions being fed in 1839. The conflict had a personal element. On 29 April 1836 Macready, the actor, punched and threatened to kill Bunn. The literary establishment--'the Shaksperian clique, by which are meant Mr. Macready and his toadies' as Bunn called them--closed ranks. Macready's supporters included Bulwer-Lytton, Carlyle, Dickens ('who adopts the facetious coxcombry of calling himself Boz'[14]), and Talfourd, all eminent authors. Talfourd, a barrister, defended Macready in the ensuing court case. Never would these contribute libretti to the mainstream of English opera at Drury Lane.

Bunn later antagonized another group of dramatists. His second period at Drury Lane coincided with the rise of *Punch*. Two of the staff, Mark Lemon and Gilbert à Beckett, had operatic experience. Lemon had collaborated with his brother-in-law, Romer, in *The Pacha's Bridal* (1838) and *Fridolin* (1840), and with Loder (*The Deer Stalker*, 1841). A Beckett was the author of *Agnes Sorel* (1835), set by his wife, Mary--a unique case of wife as composer and husband as librettist--and the translator of Balfe's French operas *Le puits d'amour* and *Les quatre fils Aymon* for performances at the Princess's in 1843 and 1844. He ran his own *Almanack of the Month* in 1846, satirizing Jullien and Lablache, laughing at rondo finales, praising Loder's *The Night Dancers*, first performed earlier that year, and hinting at the back-biting prevalent in the British musical scene. But he reserved most of his invective for Bunn for not supporting untried composers, and included a skit on a *Maritana*-type plot (*Maritana*, with text by Fitzball and music by Wallace, appeared at Drury Lane in 1845).

This was gentle compared to *Punch's* satire. Having initially been treated with bonhomie, by 1845 Bunn was regularly lampooned as 'the Poet Bunn', and his ballads were mercilessly dissected. The ballad he wrote at Fitzball's request in *Maritana*, 'In happy moments, day by day,/The sands of life may pass', was treated as follows:

> This idea was probably suggested to the Poet BUNN by the daily passing of a sand-cart before his door ... we ... offer one pound (of gingerbread) reward to any charity boy in any of the metropolitan parishes, who will produce a more complicated piece of nonsense than this, which, "at the request of the Author and Composer," the Poet BUNN has perpetrated.[15]

Occasionally 'the Poet's own musician, BALFE' was brought into the fray. Bunn restrained himself until he was no longer the lessee of Drury Lane. He then produced *A Word with Punch* [1847], a brilliant spoof, backed up with acerbic pictorial caricatures, and attacking the personalities with their own weapons. 'Unlike the other two,' Douglas Jerrold, a playwright and contributor to *Punch*, was 'a man of undoubted genius.... He is one of the most ill-conditioned, spiteful, vindictive, and venomous writers in existence.... it never once entered my head to ask your puppets to write an opera for me, being perfectly convinced, despite their opinion, that they have not one requisite for the undertaking.' No composer ever 'wed "exquisite music" to *their* words.' Bunn then dissected Lemon's libretto ('I'll be bound you never read *Thickhead's* opera of *Fridolin*') with predictable glee.[16] The British Library's copy bears a note that this 'was too true to be pleasant, and has been industriously bought up, and destroyed, so that it is now very scarce.'

Eventually Bunn was defeated by 'the conclave of Infernal Fools & of Satanic Scoundrels who have succeeded by spending 100,000 £ [*sic*] in injuring me so seriously that I no longer care personally which way the matter goes.'[17] The issue reveals a sectarian world in which the most inventive literary minds were deeply divided from the world of opera. When Bunn died of apoplexy in Boulogne on 20 December 1860, Pyne and Harrison were cancelling Balfe's *Bianca, the Bravo's Bride* to make room for the lucrative Christmas pantomime. Chorley squeezed the following obituary into the *Athenaeum*'s last issue of the year:

> The death of Mr. Bunn, the well-known manager, playwright, and author, which took place suddenly, at Boulogne, last week, claims a line here. He was active and energetic, but unscrupulous and unwise in all the three characters; and thus, though hard working, and, after his coarse fashion, zealous to catch "the town," he did not succeed, because he did not deserve to do so.

Balfe told Planché he had visited Bunn's grave. 'Well, never mind, poor Alfred!' said Balfe; and then, quoting Bunn's own words in *The Daughter of St Mark*, "We may be happy yet." The line comes from Adolphe's ballad 'Oh, smile as thou wert wont to smile,/Before the weight of care/Had crushed thy heart.'[18]

Bunn's animosities with the Bulwer-Macready faction and the *Punch* group ruled out many potential librettists. Certain composers--Fitzwilliam, Glover, Linley, Loder, Lover, and Rodwell--wrote some of their own libretti, and Macfarren senior wrote for his son. Mary Russell Mitford, Dickens, Chorley, and Boucicault, authors better known in other fields, contributed one libretto apiece. With the exception of Bunn, however, most librettists were professional men of the theatre who wrote whatever range of theatrical works was currently in demand, and often had other personal and business interests. Several produced only a single libretto. Soane, for instance, who had written *Aladdin* for Bishop in 1826, supplied various translations, but one original libretto (Loder's *The Night Dancers*, twenty years later) during this period. Likewise Planché, the author of Weber's *Oberon* (1826) and one of the most productive of theatrical writers, spent more time on other genres, including translations: his only libretto between 1834 and 1864 was *Love's Triumph*, for Wallace, in 1862.

By contrast, Bunn's output was almost exclusively operatic. He wrote, wholly or in part, eleven libretti between 1836 and 1852--seven for Balfe. The complaints that he exercised a monopoly had some foundation. Fitzball, always close to the administrations of the main theatres, was even more prolific. In addition to countless other stage works, he produced fifteen libretti in the thirty-odd years under review.

Bunn's contribution to the English libretto is as much a compound as the man himself. Dramatically, much of the writing is superficial, rarely exploiting the potential of a situation, and weak in characterization. The French biographer, Pougin, castigated *Matilda of Hungary*, which Bunn wrote for Wallace, as 'lourd, trivial, sans intérêt et du plus mauvais goût.'[19] On occasion the sense becomes impenetrable or farcical. The style is often heavy-footed, self-consciously seeking rhymes, and rarely fluent, as the following heroic couplets, an example of Bunn's recitative from Act III of *The Daughter of St Mark*, illustrate:

> No more then, and begone! upon the brink
> Of fate as now she stands, if I could think
> Thy--mine no longer--Catarina, free
> *From* crime, tho' now *with* crime accused she be,
> A load were off my heart; but, 'tis in vain,
> And mine the grief, the madness, and the stain!

On the other hand, there is the beginning of a psychological exploration in the character of Moncenigo [*sic*] in the same opera; and, as we shall see in chapter 2, Bunn did establish a particular style of ballad that contributed greatly to the popularity of Balfe's and Wallace's operas. It had become traditional to denigrate him even before the days of *Punch*. A rare exception came from George Linley, himself a librettist, who defended Bunn's work in two diatribes of the 1860s that strikingly reveal the hatreds underlying the world of English opera. Linley held up Chorley's translations and libretto to scorn. 'Compared with Chorley, it is *no* mis-nomer,/'Tis *not* pseudonymous to call Bunn--Homer!'[20] This was overcompensation with a vengeance. The most balanced judgement comes from Bernard Shaw, the only critic to take into account that essential criterion for a libretto--how well it can be set and sung:

> The authorship of the inimitable Bunn is conspicuous in the singableness, the sentiment, and the outrageous absurdity of the lyrics. He never wrote words quite so pleasant to sing, so melting to hear, so irresistibly funny to read as those of When other lips and other hearts, but Let me like a soldier fall is by no means unworthy of him. A literary man who is not musical may always be detected by his inability to perceive the least merit in Bunn. Musicians know better, and envy Balfe and Wallace their librettist.[21]

A new generation of librettists came to the fore in the 1860s. The change is palpable. Compared to Fitzball's pandering to the lowest common denominator (especially evident in his patriotic vein) and to the standard with which Bunn was satisfied, Chorley's *The Amber Witch*, set by Wallace and first performed in 1861, and *The Armourer of Nantes*, which Bridgeman wrote for Balfe (1863), are distinctly superior. The libretto of Benedict's *The Lily of Killarney*, of 1862, is also of a generally high quality. It was adapted by Dion Boucicault, primarily a playwright and actor, from his play *The Colleen Bawn* (1860), itself based on Gerald Griffin's novel *The Collegians*, of 1829. Boucicault was assisted by John Oxenford, probably brought in by Pyne and Harrison to lend a more experienced hand where opera was concerned, such as in the lyrical sections. (Collaborative work on a libretto was infrequent, much less common than in Paris.) A prominent man in the theatrical world--he was the dramatic critic of *The Times* from about 1850 for over twenty-five years--Oxenford numbered nine libretti among his sixty-eight stage works, including three operas (plus the masque *Freya's Gift*) written for Macfarren between 1860 and 1864. He had a study of Schopenhauer to his credit, and was a noted translator, competent in several languages. Not surprisingly, his libretti have an intellectual stringency all too lacking in Fitzball's, for instance. But by nature he was a 'literary man who is not musical,' to borrow Shaw's words, and he could be insensitive to the issue of matching of verbal metre and musical stress. A case in point is Hardress's 'A bachelor's life is gay, careless, and free,' in the introductory chorus of *The Lily of Killarney*, where the consecutive accented syllables ('*gay*, *care*-less') make an effective setting virtually impossible.[22] A note in the printed libretto states that the words of Eily's solo 'In my wild mountain valley he sought me' (Act I), were written by neither Boucicault nor Oxenford but one 'Francis W. Brady, Esq., of Dublin'; Benedict was able to turn the variety of verbal accentuation to good account.

With the exception of Macfarren's *Robin Hood* (1860) and Wallace's *The Amber Witch*, both given at Her Majesty's while the theatre was under a short-lived management that allowed opera in English, and some smaller works by Macfarren and Balfe produced at the Marylebone Theatre and the Royal Gallery of Illustration, the majority of English operas produced between 1857 and 1864 were given by Pyne and Harrison's company, mostly at Covent Garden. Their partnership made an opera house available to English composers after a long fallow period. The company included two singers of unusual quality: Louisa Pyne herself, and the redoubtable Charles Santley. They created roles that none of their English contemporaries could attempt satisfactorily. Elvira in Balfe's *The Rose of Castille* and Danny Mann in *The Lily of Killarney* are examples. The latter, among the best parts in the English repertory, would certainly not have been written were it not for Santley. But there were many rows. Santley delighted in humiliating Harrison. In 1861 he joined a rival company, including Sims Reeves and conducted by Hallé, at Her Majesty's. During the following year he was engaged concurrently at both theatres, Her Majesty's now being under Mapleson's management. He then transferred entirely to the latter, singing Valentine in *Faust* to Gounod's complete satisfaction; the composer wrote 'Even bravest heart may swell' for him in the next season. His memoirs, *Student and Singer* (London, 1893), are a valuable source of information.

After Pyne and Harrison concluded their partnership, the composer and conductor Alfred Mellon continued the enterprise at Covent Garden for two more seasons, until 1866. The demise of the company was due to various causes, among which was the important factor of royal support. Queen Victoria, an enthusiastic theatre-goer from her youth, and Prince Albert attended many of Pyne and Harrison's performances. Between 15 February and 6 March 1859 the Queen visited Covent Garden on five occasions, seeing Balfe's *Satanella* three times in ten days and *Maritana* once. In 1860 she attended Wallace's *Lurline* twice within little more than a fortnight of its opening, and early in 1861 was present at performances of *Bianca, the Bravo's Bride*. At Her Majesty's she saw *Robin Hood* and, on 9 March, *The Amber Witch*. This was the last opera she saw in an opera house. She made her last visit to any theatre a few days later. Her mother died shortly afterwards. The Court went into mourning. Soon the Prince Consort fell ill, and he died in December. His death removed what could have been long and powerful support for English opera. In the words of Luigi Arditi, Mapleson's conductor at Her Majesty's, 'the web of powerful interest which inevitably folded itself round the opera nights' was no longer there.[23]

NOTES

1. See Benjamin Lumley, *Reminiscences of the Opera* (London, 1864), p.35.

2. [Hermann Pückler-Muskau], *Tour in Germany, Holland and England in the Years 1826, 1827, & 1828* [vols 3 and 4 of *Tour in England, Ireland, and France in the Years 1828 & 1829*, 'by a German Prince'], 4 vols (London, 1832), 4:342; 3:128-29; Sir Walter Scott, 'Essay on the Drama,' in *Essays on Chivalry, Romance, and the Drama* (London, [1888]), p.226.

3. Pückler-Muskau, 4:389.

4. Lumley, p.2.

5. Alfred Bunn, *The Stage: both before and behind the curtain*, 3 vols (London, 1840), 1:34-35.

6. While not strictly relevant here, it may be mentioned that Her Majesty's was subject to 'an almost inextricable tangle' and 'hopeless confusion'--the auditorium and the stage came under two different leases. (See Lumley, p.2.)

7. See *Survey of London*, general editor F.H.W.Sheppard, 41 vols (London, 1900-83), 35, *The Theatre Royal Drury Lane and The Royal Opera House Covent Garden*, *passim*; Bunn, *The Stage: both before and behind the curtain*, *passim*; Charles Lamb Kenney, *A Memoir of Michael William Balfe* (London, 1875), p.139; G.G.Urwin, 'Bunn and his influence in the theatre' (unpublished Ph.D. dissertation, University of London, 1956), *passim*; Urwin, 'Alfred Bunn 1796-1860: A Revaluation,' *Theatre Notebook*, 11 (1957): 96-102. Precise figures and dates vary slightly from one account to another.

8. Hector Berlioz, *The Memoirs of Hector Berlioz*, translated and edited by David Cairns (London, 1977), pp.448, 450 fn8; James Robinson Planché, *Recollections and Reflections*, 2 vols (London, 1872), 2:[132]-34.

9. Planché, 1:269-72. In 1860 Planché supplied the basis of an Italian version of *Oberon*, given at Her Majesty's with additional music by Benedict.

10. Planché, 2:200; 1:251; British Library, Additional MS 43382, f.154 (letter to Morris Barnett, tentatively dated 1846-47).

11. *The Times*, 29 May 1837, p.3.

12. BL, Add.43382, f.10 (letter from Charles Kemble to Bunn, c.1839).

13. *The Times*, 10 November 1846, p.5.

14. Bunn, 3:113; 1:6.

15. *Punch, or the London Charivari* [successively edited by Henry Mayhew and Mark Lemon, by Shirley Brooks, by Tom Taylor, by F.C.Burnand, and by Owen Seaman], (London, 1841, *etc.*), 9 (1845): 229. See also 7 (1844): 152, 271; 9 (1845): 87.

16. Alfred Bunn, *A Word with Punch* (London, [1847]), pp.4, 5, 8.

17. Letter to Morris Barnett, dated 4 March 1849. BL, Add.43382, f.68.

18. *Athenaeum*, no.1731, 29 December 1860, p.916; Planché, 2:201.

19. Arthur Pougin, *William-Vincent Wallace: étude biographique et critique* (Paris, 1866), p.15.

20. George Linley, *Musical Cynics of London: A Satire (Sketch the First)* (London, 1862); *The Modern Hudibras* (London, 1864). The quotation is from *Musical Cynics*, p.12.

21. Bernard Shaw, *Shaw's Music*, 2nd rev. edition by Dan. H. Laurence, 3 vols (London, 1989), 1:226. 'When other lips' is from *The Bohemian Girl*, 'Yes, let me like a soldier fall' from *Maritana*.

22. Similar defects occur in his translation of *Rigoletto* for Boosey's edition (1871)--for example, '*I* am *con*-ten-ted' for 'rin-grazia-to' (p.61 of the vocal score), or '*Wo*-men are | *fi*ckle AND | ' for 'La donna è mobile.' In Novello's rival edition of 1872, Natalia Macfarren, the composer's wife, rendered the first instance as 'thanks are *gi*-ven,' and, in common with most other English translators, did not attempt a literal version of the Duke's aria: '*Plume* in the | *sum*-mer wind.' In the translation of Masse's *La Reine Topaze* (1860), another collaborative venture, Oxenford left the lyrical portions to George Linley, a composer as well as a librettist. (*The British Library General Catalogue of Printed Books* dates the publication of the libretto as '[?1865],' but the title page specifies performances in 1860.)

23. Luigi Arditi, *My Reminiscences* (London, 1896), p.96. See also Willert Beale, *The Light of Other Days*, 2 vols (London, 1890), 2:232-33; George Rowell, *Queen Victoria Goes to the Theatre* (London, 1978), pp.137-38.

II

The Conventions of the English Opera Libretto: Plot and Character

The major feature of the English opera libretto in the nineteenth century is its particular type of plot. The task of the present chapter is to identify and offer an explanation for its conventions, including the important types of characters.

It would be a reasonable expectation that, in the land of Shakespeare and other eminent authors, a proportion of national operas was based directly or indirectly on the famous dramatic canon. A close study of all known libretti of the period reveals that this was not so. Tragedy was regarded as the exclusive province of the spoken theatre. No English opera ends unhappily. There is no native operatic setting of *Othello*, for instance. Instead, the libretto was constructed according to certain procedures that were observed without exception. The clearest means of establishing these is to turn to the most popular opera of the entire genre, Balfe's greatest success--*The Bohemian Girl*, of 1843.

In nineteenth-century terms, the action took place fairly recently, for one character is the Austrian governor of Presburg and another a proscribed Pole. The basic events are as follows. The governor's six-years old daughter is kidnapped. Twelve years elapse, during which she is brought up by gipsies. She retains no conscious memories of her past. Also in the gipsy community is the Pole, a political fugitive, and they fall in love. This brings upon her the hatred of the gipsy queen, and on a visit to Presburg she is brought to trial on a charge trumped up by her rival. Vainly she protests her innocence until, unable to accept the injustice of the likely outcome, she tries to commit suicide. She is on the point of stabbing herself when her judge, the governor, notices a mark on her arm that identifies her as his long-lost daughter. The truth is revealed. With her rage now directed at the man who rejected her, the gipsy queen orders one of her men to shoot the Pole; but a friend deflects the weapon and the queen is killed by the very bullet she intended for the hero. The work ends with general rejoicing, heightened by the prospect of the marriage of the heroine and the hero, who is now warmly accepted into the community.

The central character is the heroine, whom we first meet as an innocent child. She develops from child to adult through three clear stages: the initial environment of family; the second phase in which she matures to adulthood, confronting simultaneously the constructive and destructive power of adult passions; and, finally, not so much a return to her original environment as a reaffirmation, a generation later, of its tenets. Her sense of morality is important. Even when unaware of her own nominal identity, she loses none of her childhood virtue. It is this that is tried to the utmost, but by not wavering she brings about the overthrow of evil. Indeed, her identity depends on it. At the outset, the heroine is at the bosom of one family; in due course, we may imagine, she will be at the centre of another.

There are two main instances of evil, the kidnap and the trial. One is successful, the other very nearly so. We know nothing of the early backgrounds of its proponents. If we assume that they, too, were once imbued with childhood innocence, it has been overcome by other forces. The kidnap was motivated by vengeance. The second attack was more complex. It was caused by revenge and hate--passions destructive of individuals and of family.

The manichaeistic conflict and its outcome lie at the heart of the English opera libretto. Its essence is melodrama. In this regard, libretti are indistinguishable from many other theatrical works of the day, for such plots were immensely popular, far more so than their loftier dramatic counterparts. The libretto was as much a reflection of contemporary taste as the hundreds of works that followed Holcroft's *A Tale of Mystery* (1802), itself an adaptation of Pixerécourt's *Coelina; ou, l'enfant du mystère*, produced in Paris two years earlier.

Melodrama has become a pejorative term for excessive gestures unsupported by an adequate plot, of effects exceeding causes, of exploitation of our responses to ephemeral stimuli. However, the extraordinary place it held in the nineteenth-century mentality requires an unbiased approach. An understanding of melodrama is essential if we are to understand crucial aspects of nineteenth-century English opera.

We must be clear about the use of the term, which has several connotations. One usage denotes sections of stage works that are delivered in speech accompanied by music. Another application of the term refers to a particular repertory of stage works--Pixerécourt's French mélodrames, for instance (themselves influenced by English gothic literature). In this study 'melodrama' is used in a broad, conceptual sense to refer to a genre that embodied particular issues and attitudes, not necessarily in a particular medium, characteristic of much European culture of the late eighteenth and the nineteenth centuries. Peter Brooks has referred to melodrama as 'a mode of conception and expression, as a certain fictional system for making sense of experience, as a semantic field of force.'[1]

English opera was one of the media, as was the mélodrame, whether staged in Paris or in London. So, in large measure, was the novel, whose development was a special feature of the period. They were closely related. Novelists based episodes in their tales on specific stage works, and dramatists adapted novels--in the case of Dickens's and Ainsworth's, even before the novelist's serialized version was complete. The relationship between the opera libretto and the mélodrame of the Pixerécourt type is shown by various examples. An operatic work could be performed as a mélodrame, simply by removing the majority of its music, as in the case of Rodwell's *The Sexton of Cologne* in 1837.[2] An early staged version of Griffin's novel *The Collegians*, prior to Boucicault's *The Colleen Bawn* and the opera *The Lily of Killarney*, provides another instance of the common, international, thread underlying so many works. Stirling's spoken 'drama,' *Aline: or, The Rose of Killarney* (1843), though set in Ireland, has only brief references to the novel. In fact it was a rival version with C.Z.Barnett's *Linda, the Pearl of Savoy* of d'Ennery and Lemoine's drama 'mêlé de chant' *La Grace de Dieu* (1841), which had been heard in yet another version a few weeks earlier at Her Majesty's--Donizetti's *Linda di Chamouni* (1842). Ireland served as Britain's equivalent to picturesque Switzerland.

To explain the extraordinary grasp of the melodramatic concept, we need to be aware of broader cultural issues before and during the period now studied, each of which had a bearing on the matter.

The assumption that one could gainfully view the world in terms of a clear conflict between virtue (usually equated in libretti with innocence) and evil, each presented in unambiguous terms, was widely upheld. It had been stated by important authorities in the eighteenth century. The author of the French *Encyclopédie*'s article on virtue associated it with a constant effort to maintain God's laws despite human frailty--an unremitting spiritual exercise to overcome the inherent weaknesses of humanity. The role of art and artists in this task was clearly stated. Samuel Richardson had made his position clear in 1740, with *Pamela*'s subtitle--*Virtue rewarded.... Now first published in order to cultivate the principles of virtue and religion in the minds of the youth of both sexes*. Dr Johnson insisted on the representation of

> the most perfect Idea of Virtue; of Virtue not angelical, nor above Probability; for what we cannot credit we shall never imitate; but of the highest and purest Kind that Humanity can reach, which, when exercised in such Trials as the various Revolutions of Things shall bring upon it, may, by conquering some Calamities, and enduring others, teach us what we may hope, and what we can perform. Vice, for Vice is necessary to be shewn, should always disgust.[3]

In France, Diderot defined the responsibility of the artist as 'Rendre la vertu aimable, le vice odieux, le ridicule saillant, voilà le projet de tout honnête-homme qui prend la plume, le pinceau, ou le ciseau.'[4] Greuze's paintings were examples of the 'exemplum virtutis' in the visual arts. By the end of the century this intention was an article of faith, as Mrs Radcliffe's concluding words in her famous gothic novel *The Mysteries of Udolpho*, of 1794, illustrate.

> O! useful may it be to have shewn, that, though the vicious can sometimes pour affliction upon the good, their power is transient and their punishment certain; and that innocence, though oppressed by injustice, shall, supported by patience, finally triumph over misfortune!

Almost fifty years later, in 1840, the *Athenaeum* interpreted an illustration in Dickens's *The Old Curiosity Shop* in similar terms: 'in the midst of Violence, Superstition, and all the hateful or hurtful Passions of the World ... [h]ow soothing the moral, that Gentleness, Purity, and Truth, sometimes dormant but never dead, have survived, and will outlive, Fraud and Force, though backed by gold and encased in steel!'[5]

These sentiments also underlie the libretto. Normally there was no opportunity for librettists to make such a clear statement, but Planché's explanation of *The Magic Flute*, printed in the libretto on the occasion of its first performance in English in 1838, illustrates the same mode of thinking. (The part of Papageno, heavily reduced by Planché, was sung by Balfe.) It is a manichaeistic conflict between 'reason or light' and 'the passions without reason, or darkness; and the whole plot of this Opera turns upon the struggle between these two oldest of contending parties'. The 'Magic Flute' itself

> has the power of inspiring Love, the most potent of human passions. Bestowed ... by the powers of darkness and evil, it is of course merely sensual--purified by the powers of light and reason its magic is made subservient to the best and holiest of purposes, and guides the faithful pair through all worldly dangers to the knowledge of heavenly TRUTH.

The religious element to which Planché alludes, already evident in Richardson's *Pamela*, was strong. In the early years of the nineteenth century, Byron's friend, R.C. Dallas, on whose translation of Madame de Genlis's novel *La Siège de la Rochelle* Balfe's *The Siege of Rochelle* was based, substituted 'or, the Christian heroine' for Mme de Genlis's original subtitle 'ou le Malheur et la conscience.' By the mid nineteenth century the connection was even more firmly established. It reached a particular point, for instance, in George Eliot's tale of 'Janet's Repentance,' in which the moral is delivered by a reformed transgressor who has taken the cloth. The melodramatic ritual is not a complete parallel to Christian belief, for it lacks the essential element of Redemption, but its Christian outlook was another reason for its popularity. The Platonic theory of a perfectly virtuous being, advocated by Dr Johnson and others, explains the libretto's concept of characters. Today, we tend to judge a work by the depth of its psychological portrayal of individuals, often revealed by a degree of inner change in their psychological conditions. We must not expect this in the libretto. *The Bohemian Girl*, for instance, does not offer a case-study of the traumatic effects on a young girl of kidnap, upbringing in an alien community, and desperation that drives her to the point of suicide. We learn nothing about the heroine's inner self in this sense. The stock heroine does not grow. She is a static, idealized version of Woman: always virtuous, as unchanging as are the vast majority of other stock characters. Rarely are there less monolithic characters. This, in twentieth-century terms, is the flaw in the genre: the figures moving about the stage are mere representative puppets, gesturing their way through an elemental conflict. The nineteenth century adopted a different view. Audiences wanted not investigations of the human psyche but portrayals of humans as emblems of fundamental forces. The characters' lack of distinctive personal traits made identification with their moral qualities easier.

On the basis of the above, we may now consider the three main protagonists of the conventional conflict--the heroine, the villain, and the hero.

THE HEROINE

By the 1830s there was a long tradition of virtuous heroines extending well into the previous century. Sally, in Arne's *Thomas and Sally* (1760), enunciates the basic creed of all her successors:

> Grant me, ye pow'rs, I ask not wealth;
> Grant me but innocence and health;
> Ah! what is grandeur link'd to vice?
> 'Tis only virtue gives it price.

Clari, Bishop's 'Maid of Milan,' learns the wisdom of Sally's words through bitter experience. Clari is one of the rare instances of a fallen woman as heroine in English opera. She does, of course, reform, and returns to the 'Home, sweet home' of her Noble-Peasant father. The only example in our period of a heroine whose virtue has weakened is Marie in Balfe's *The Armourer of Nantes* of 1863, deliberately made ambiguous by Bridgeman compared to the character in Victor Hugo's *Marie Tudor* (1833), on which he based his libretto, and to Ghislanzoni's Italian version, *Maria Tudor*, set by Kashperov (1859) and later by Gomes (1879). Other libretti deal with the issue by awakening the heroine from a dream, as in Mellon's *Victorine* of 1859, leaving her pure after a sobering experience.

One of the few changes in the heroine up to the 1860s was the weakening of the Rousseauesque influence evident in *Clari*. Later figures tend to be urban and middle- or upper-class. The Irish peasant-girl Eily in *The Lily of Killarney* is a late exception. An unchanging feature was the attitude to insanity. No English operatic heroine goes mad, for the same reason that emerges in Charlotte Bronte's *Jane Eyre*

(1847). The dissolute Bertha Rochester does disintegrate, but the pure Jane Eyre does not: she is sustained by virtue.

This is the crucial difference between the villain and the heroine. She never loses the innocence of her childhood. The childlike quality can influence her relationship with the operatic hero, who is frequently older and to some extent takes a parental role. Innocence is so important that, if threatened, the heroine is willing to sacrifice her life to preserve it. Arline in *The Bohemian Girl* is one example, and the heroine of Wallace's *Maritana* another: 'A captive, in a stranger's power,/ She'll perish ere she yield to shame!' she cries. (Mrs Davidson's contemporary translation of d'Ennery and Dumanoir's play *Don César de Bazan*, on which *Maritana* was based, was more explicit.) The eponymous heroine of Lavenu's *Loretta* (1846), for which Bunn wrote the libretto, tries to commit suicide after being raped.[6]

The manichaeistic plot requires the heroine to be caught in dilemmas that would crush a person of lesser fibre. Playbills trumpet her moments of greatest trial. The clearest instance of which I am aware does not strictly fall within the realm of English opera, but is worth quoting: the bills for *Quasimodo*, adapted by Fitzball from Hugo and with music selected by Rodwell from Mercadante, Beethoven, and Weber (1836), promised a 'PROCESSION CONDUCTING ESMERALDA TO THE SCAFFOLD!... the SCAFFOLD, erected for the EXECUTION of ESMERALDA!' The heroine is always alone in a male world; her mother is dead. (The one exception to the rule is in *The Siege of Rochelle*, but the heroine is unaware of this.) Her physical weakness adds to the tension.

At such moments she appeals to the highest power. 'To heaven I leave the vindication of my innocence: its justice is sure, though sometimes slow, and the hour of retribution often arrives, when least expected'. So speaks Amanda in Mrs Roche's *The Children of the Abbey*, a novel from the late eighteenth century whose fame continued well into the nineteenth.[7] Clara, the heroine of Fitzball's *The Siege of Rochelle*, echoes Amanda's words--'Heav'n, which knows I am not guilty,/ In its time will send redress'--as do Arline and Catarina, the heroines of Bunn's *The Bohemian Girl* and *The Daughter of St Mark*, each of whom, in the extremes of anguish, cry 'Heaven knows I am innocent.' Since this appeal is called forth by the very success that the villain now enjoys, the heroine's triumph--the triumph of virtue--is made all the more satisfying when it does arrive. Without such displays of courage, the figure of the heroine would risk becoming a mere cipher for virtuous, dutiful behaviour. Her deference to her father, for instance, is extreme. *The Siege of Rochelle* depends upon the heroine obeying the man she believes to be her father though her own life will suffer. In Barnett's *Fair Rosamond* of 1837 and in two operas of 1860, Wallace's *Lurline* and Macfarren's *Robin Hood*, the heroine pleads with her father, but only in *Bianca, the Bravo's Bride*, by Balfe (also dating from 1860), does she fully challenge his authority. In each case she is motivated by love for the hero.

In an age when equations were drawn between the internal and the external, according to one theory (Lavater's) the heroine's beauty itself reinforced her virtue.[8] In *The Bohemian Girl* Arline's beauty is one reason why her judge cannot believe she could be guilty of robbery. The cult of the heroine was further strengthened by a small number of individuals who gained an unassailable place in the public's affection. Among these were certain ballerinas. Marie Taglioni, Fanny Elssler, Carlotta Grisi (who created the role of *Giselle* in Paris in 1841), Fanny Cerito, and Lucile Grahn all became legendary. London audiences were no less fanatical than their Parisian counterparts. Thackeray obviously expected the readers of *Vanity Fair* (1848) to be familiar with Taglioni's name.[9] Two other women, from different walks of life, contributed to the cult. Jenny Lind, the 'Swedish Nightingale,' owed her fame not only to her voice. She was revered as the epitome of the unsullied, virtuous heroine, deeply religious, 'whose purity and simplicity won all hearts.' Writing in the 1890s, her biographers Holland and Rockstro emphasized this point over and over again,[10] dwelling on the exemplary way she retained a proper sense of values when surrounded by kings and princes. Their words read like the description of an operatic heroine. When Alfred Bunn took the extraordinary decision to put her on trial for breaking her contract, he was casting himself as the villain in his own plot. Lind never did appear in English opera. Had her predecessor, Malibran, who championed Balfe's works, not died at an early age, and had Lind sung at Drury Lane, the history of English opera in the nineteenth century might have been very different.

The images of the heroines played by the ballerinas, by Jenny Lind, and by other favourite performers left a deep mark on many a young girl's heart. One such was the person who had found herself the heroine of a melodrama played out on a truly national stage: Queen Victoria. Victoria loved the theatre, and her personal papers attest to the admiration she felt for the stage heroines and her loyalty to the performers. In this she was probably no different from any other young woman. But her position gave her special status. The fatherless, innocent child, thrust at the age of eighteen (just the right melodramatic age) into grave circumstances from which she emerged unbowed, thereafter progressing to marriage ana family with all the success of an operatic heroine--this must have been a popular perception.

As we have seen in the case of *The Bohemian Girl*, English opera strongly advocated the family as an institution. Lionel Trilling and Harold Bloom have described the latter as 'the palladium of English piety.' 'No sentiment,' they point out, 'had a more tenacious hold upon the English than that family life was the crown of personal existence and the source of personal and national virtue.'[11] Novels and paintings articulated the belief in various ways. Dickens's *The Christmas Carol* (1843) and Augustus Egg's *Past and Present* (1858) are examples. It was a strongly middle-class feeling. Regarding this point, we may note that, unlike the most celebrated heroine of the previous generation, Bishop's eponymous Clari, Arline, the outstanding heroine of the 1830s to 1860s, is not the daughter of a Noble Peasant. Equally, in her famous 'Dream,' the heroine rejects the aristocratic society of knights and nobility for the love of a man whose social standing she discovers much later. On the rare occasions when the hero fell short of the ideal, as with Rudolph in *Lurline*, it was by demonstrating her own values that the heroine would reclaim him. In such instances, the woman's role was to bring her beloved back to the paths of righteousness when he was tempted to stray. By 1837 the Count's effusion in Rooke's *Amilie, or, The Love Test* was outdatedly purple (the work had been composed twenty years earlier), but it gives us a very clear understanding of the heroine's status:

> Oh, nature, wond'rous mother, thou hast given
> To man foretaste of heaven, in woman's love;
> Firm as the rock, yet meek and lovely pure; [*sic*]
> Nor joy nor sorrow e'er can change its truth.

THE VILLAIN

Unlike their virtuous counterparts, villains explained themselves in detail. Their two main motivations were revenge and hate caused by rejection in love, and lust for wealth and power. Other motivations were extremely rare. One occurs in the early part of the period, in Benedict's *The Gipsy's Warning*, where the villain's lieutenant, Gasparro, has been twisted by his ill-treatment at the hands of the Inquisition, but he is not a central character, and he is reformed by the end of the opera. Bunn contributed one extraordinary instance, in Lavenu's *Loretta* (1846), where a hedonistic nobleman rapes the heroine. He is immediately overcome by remorse and eventually marries her. This is the only libretto in the entire period to deal with such a crime. Against all one's expectations, *The Times* passed no special comment in its review. The only remaining motivation is Danny Mann's mistaken loyalty to his master in *The Lily of Killarney*, which leads him to attempt murder--an outstanding case, to which we shall turn in due course.

Danny's role is a mark of the considerable increase in subtlety of characterization that emerged in the final phase of the thirty-odd years now under review. In reading early libretti one is sometimes reminded of Mrs Radcliffe's didacticism in *The Mysteries of Udolpho*: the villain's soliloquy is a case-study of the fall into evil. The development in approach took place gradually. It required a more sophisticated portrayal of human nature on the part of librettists, and involved no more than a handful of works.

However much more subtle their manner, even the later librettists subscribed to an important outlook concerning the evil personality: what we may anachronistically call contemporary psychology. Especially in the early stages, librettists used a vocabulary that can now appear bombastic and weak if we are unaware of its context and meanings. They drew upon a long and honoured body of thought, the study of the passions. It will be useful to consider this subject before we turn to particular examples of evil characters, for it influenced the styles of texts and of acting.

Quotations found in chapter 1 of this study refer to the passions. The concept was widely accepted in Britain and abroad. It involved an 'analysis of human nature,' as Lord Kames had called it in his *Elements of Criticism*,[12] founded on the categorization of fundamental emotional impulses--Anger, Fear, Love, Sorrow, and so on. The Baroque aria, with its concentration on Affections, owed much to the notion. Investigations ranged from Descartes's *Les Passions de l'Âme* (1649) to a long article in the *Encyclopédie* (1765), Charles Darwin's *The Expression of the Emotions in Man and Animals* (1872), and William

Archer's *Masks or Faces? A Study in the Psychology of Acting* (1888). By 1802 R.C.Dallas could take such interest for granted, whereas the study of anatomy needed special pleading. The 'Analysis of the Passions' in his *Elements of Self-Knowledge* emphasizes the care and precision with which the passions were considered (2.1).

Passions were expressed in standardized gestures that were traced to Aristotle and Quintillian. Dickens's sketch of a 'theatrical young gentleman ... [who] looks upon Mr. Fitz Ball [*sic*] as the principal dramatic genius and poet of the day' tells us how widely understood these conventional gestures were. He

> is ... very acute in judging of natural expressions of the passions, and knows precisely the frown, wink, nod, or leer, which stands for any one of them, or the means by which it may be converted into any other.... If you venture to express a doubt of the accuracy of any of these portraitures, the theatrical young gentleman assures you, with a haughty smile, that it always has been done in that way, and he supposes they are not going to change it at this time of day to please you.[13]

The appearance of the heroine of *Clari, or The Maid of Milan* in 1823--'her hair disshevelled [*sic*], her face pale'--was the conventional signal for distraction and grief. Such gestures crossed borders of genre and nationality. In *The Bohemian Girl*, for instance, Bunn translated instructions in St Georges's and Mazillier's ballet-pantomime *La Gypsy* almost verbatim: compare 'une femme pâle, échevelée, s'y montre avec les gestes de la plus affreuse terreur' with 'BUDA, *pale, and with dishevelled hair, signifies by her gestures that* ARLINE *has disappeared.*' In *Loretta* Bunn was unusually expansive, probably because the part of the heroine was performed by the international soprano Anna Bishop. The rapist, Don Carlos, is soon horrified at his actions--'*pale and distracted in his look and manner.*' Loretta is completely traumatized. She '*has scarcely strength to sustain herself; her hair is dishevelled--her eyes fixed and immoveable, conveying an idea of her reason having fled.*' Some years later she finds herself in the same room. Recognition slowly dawns:

> LORETTA *sinks into an arm-chair, plunged in the deepest grief, yet struggling with her thoughts; she slightly lifts up her eyes, and looking languidly and slowly around her, a*

> *vague and confused recollection seems to come over her--*
> *she gazes more attentively, and the truth appears suddenly*
> *to flash across her mind.*

She '*clasps her forehead*' at the recognition and confronts Carlos, who '*recoils as if inwardly writhing with remorse.*' Soon he '*starts, reels back.*' As the scene closes, Carlos '*stands in a sort of stupefaction*': Loretta '*has swooned.*' Bunn used visual, not verbal, expression. How Lavenu responded we shall never know, for the music has been lost.

The influence of the passions in villains' arias is evident in the first substantial example of the period--Hela, in *The Mountain Sylph*. The origin of his malevolence is made clear: 'shafts of love' poisoned his soul. Planché's remarks on *The Magic Flute* are apposite. Hela is filled with hate. Delight in destruction has become an insatiable need, ultimately self-destructive. With its definition of the source of evil and its list of evil passions, redolent of Dallas's analytical table, the whole statement is cast in what might be called the English style. It is far more detailed than its closest German equivalent, Caspar's aria in *Der Freischütz*. The reference to his childhood innocence is noteworthy.

In days gone by
Alas, was I,
A flow'ret gay as rare,
With heart as mild
As infant child,
And features too as fair.

But doom'd to prove
The shafts of love,
From early virtues wean'd;
From good estranged
Love's poison changed
The flow'ret to a fiend.

Now Jealousy, Anger, Revenge and fell Hate,
My soul have seduced, and so mingled my fate,
From passion to passion I fly, and the slave
Of passion can only find rest in the grave![14]

Revenge was equally powerful in characters who were less overtly demonic. Queen Eleanor in Barnett's *Fair Rosamond*, Elizabeth I in Balfe's *Catherine Grey* (both 1837), and the gipsy queen in *The Bohemian Girl* (1843) are all led to the point of murder by this passion. *Catherine Grey* contained two weighty soprano roles for the heroine and the queen, and the portrait of Elizabeth I is the most expansive of the three treatments of a woman scorned. Because of its historical status, however, we shall concentrate on *The Bohemian Girl*.

Bunn did not attempt to expand the role of the gipsy queen beyond her original in the ballet on which he based the work. She is consumed by jealousy, revenge, and hate, and is enraged by the failure of her plan to remove Arline by means of a trumped-up charge. Initially she focuses her hatred on her rival, but an ominous ring enters her words:

Queen [darting a furious look at ARLINE *as she passes her.]*
Weep rivers--for ages pine!
He shall never be thine.

At last she turns her hatred on Thaddeus. Just when all seems safe,

> *as* THADDEUS *is about to embrace* ARLINE, *the* QUEEN, *in a transport of rage, points him out to a Gipsy by her side, who is in the act of firing at him, when* DEVILSHOOF, *who has tracked their steps, averts the Gipsy's aim, and by a rapid movement turns the musket towards the* QUEEN*--it goes off, and she falls.*

It is all done by wordless action. When Bunn *did* put words into the queen's mouth, as in her duet with Devilshoof, she speaks solely in the language of the passions. But the amount of text used to explain her actions is among the least in any opera of the period. She is an extreme version of the melodramatic villain: virtually an abstraction. The role of Devilshoof, her disaffected lieutenant, also betrays its balletic origin. His mixture of vindictiveness and loyalty is inadequately explained. We observe theatrical mechanisms, not causes and effects. In this regard, it must be said that Balfe's most famous work contributes more than any other to the low opinion in which melodramatic plots are held.

Bunn's ability is shown to better advantage in his portrayal of the greed and ambition of Moncenigo [*sic*], the villain in *The Daughter of St Mark*, which he founded on *La Reine de Chypre*, written by St Georges for Halévy (1841). St Georges's Mocenigo is a Machiavellian envoy of the Venetian Council of Ten. His one substantial statement--a gambling song--cloaks rather than reveals his personality. Sacchéro, who supplied Donizetti with his version (*Catarina Cornaro*), gathered St Georges's various small statements of Mocenigo's fanatical devotion to the State together in a single recitative and aria totalling eighteen lines--a celebration of 'i paterni miei lari.'

Bunn's Moncenigo is quite different. His soliloquy consists of sixteen lines of recitative--heroic couplets--and an aria consisting of two verses of thirteen lines, shedding light on the dark recesses of his personality. An appeal to 'patria mia' might have been convincing to an Italian audience, but Bunn's English approach almost eliminated this element in favour of a personal exploration that is more revealing than St Georges's or Sacchéro's character. The fundamental premise and the terminology are based on the passions. Bunn brought together a range of emotions: disdain for the ostentatious trappings of power, but delight in consolidating Venice's and his own substantial authority; awareness of the hatred directed towards him; and, most interesting, a hatred and desire for revenge, not fully understood, directed at the heroine's uncle and protector, Andréa.

> [My] object still is, wealth and station *there*,
> Without the odium *here* I'm doomed to bear--
> And then to carry out the hate I trace
> Within my breast to Andréa and his race--
> Some impulse undefined--some latent thought--
> From jealousy, suspicion, passion, caught,
> Perhaps unjustly, but imbibed so deep,
> That vengeance cannot pause, nor action sleep,
> And these two principles, in force combined,
> Excite the feeling and direct the mind.
> ....
> In revenge there's delight,
> With that feeling, at its height--
> And such revenge, in all its might,
> I revel in to-night.
> 'Tis in vain we would quell
> All the storms the heart which swell,
> Or seek to calm the griefs that dwell
> Within that secret cell....

This delving into the 'secret cell' of the personality catches attention. At the end of the opera we learn the cause of his hatred for Andréa's family. It stems from a suspicion that Andréa knew he had seduced a young noblewoman and then abandoned her when she was pregnant. Later she died. Andréa cared for her daughter, Catarina, whom the world believes to be his niece. Like various other heroines, she has been saved by a good man and brought up in ignorance of her true identity. At the denouement Andréa reveals the truth of Catarina's parentage, which Moncenigo acknowledges: 'And thus atone for years of guilt/And blood a false revenge had spilt!' He had come very close to spilling the blood of his own child, having engineered a trial that was followed by the scene of a public execution: a situation familiar from *La Juive*.

Bunn made a particular effort in *The Daughter of St Mark*--it was a special opera, being set entirely to music--but he was not sufficiently interested to sustain this level. He sank to his lowest in portraying Count Magnus in Wallace's *Matilda of Hungary*, a confused, weak role in which the Count's desire for power and his feelings for Queen Matilda are treated in a highly convoluted way.

With the exception of Brougham, whose villainous Gonzagues in *Blanche de Nevers* (set by Balfe, first performed in 1863) represents little advance on Bunn's usual standard, the next generation of librettists adopted a more thoughtful approach. Malespina in *Bianca, the Bravo's Bride*, a libretto by Simpson for Balfe (1860), Chorley's portrait of the Commandant in *The Amber Witch* (set by Wallace, 1861), and Fabio Fabiani in *The Armourer of Nantes* of 1863, written by Bridgeman for Balfe, though driven by the established motivations of lust for the heroine or for wealth and power, are drawn with deeper concern. Two characters have unusual personalities: the Baron in Loder's *Raymond and Agnes* (1855) and Danny Mann in *The Lily of Killarney*.

The terminology of the passions is still in evidence, but it is used more discreetly. Librettists were now not content merely to include established words and phrases and assume this would be adequate explanation of a character's driving-force. There is more psychological penetration. To a degree, the opera libretto was following the advances already made in other literary spheres--the novel and play.

The particular nature of the characterization and the means by which it is expressed vary according to the works on which individual operas were based and on the extent to which the librettists responded to performers' wishes. In adapting *The Armourer of Nantes* from Hugo's *Marie Tudor*, Bridgeman changed the place and period from mid sixteenth-century England to late fifteenth-century Nantes, probably to avoid any reference to an English queen and her ill-fated love for an unpopular foreigner. Clearly tact was required: Victoria had been left desolate by Prince Albert's death in 1861, two years previously, and the Prince of Wales's marriage to a foreign princess took place a month after the first night. Such changes of locale and time left the quality of Hugo's writing unimpaired. It inspired Bridgeman to produce work of a higher standard than Balfe had ever received, as the following example from Raoul's duet with Fabio shows. Raoul (Gilbert, in Hugo's play), a poor armourer who has nurtured Marie ever since she was abandoned as a baby, now loves her. Fabio has discovered that, due to her parentage, she may prevent him from gaining even more wealth. Rather than killing her outright (of which he is quite capable: he murders a Jew, the only other person aware of her heritage), he decides to destroy her reputation and thereby discredit her completely. Raoul becomes aware of his advances to Marie and a conflict with Fabio ensues. The words are Bridgeman's, not Hugo's, for this is not a direct translation from *Marie Tudor*.

Raoul

Not all thy blood shall e'er efface
The loathsome stain, the slimy trace,
    That marks thy false tongue's path.
No pow'r on earth shall e'er assuage
My bitter hate, my burning rage,
    Or screen thee from my wrath![15]

It is from such responses on the part of other characters, and from his own actions, that we learn of Fabio's evil nature--not from a soliloquy expressed in a large aria. Bridgeman did not alter Hugo's structure. Charles Santley, who took the part of Fabio, was very angry at not having an extended solo. Dramatically, however, Fabio's two large duets (with the Jew and with Raoul) both reveal his personality and avoid disrupting the continuity of the action.

Boucicault and Oxenford adopted a different structural approach in the case of Danny Mann's aria. In deference to Santley they coalesced *The Colleen Bawn*'s various short speeches and allusions into a long aria, complete with cabaletta. It was done very skilfully; and what the operatic version loses as regards the play's consistency of vocabulary (the aria's poetic English is far from pseudo-Irish brogue), it gains in concentration.

In historical terms, Danny's aria is the clearest indication of the breaking-down of the old attitude towards the passions and their influence on human motivation, earlier exemplified by Dallas's view of human nature, with its diametrical opposition of opposing emotions. Hela's character was monolithic, dominated by one branch of emotions. All others had been expunged. He was completely unrepentant.

There had been one particular instance prior to *The Lily of Killarney* in which one can sense signs of change in this regard: the Baron in *Raymond and Agnes*. Years ago the Baron killed an innocent man, abducted his wife, and left their son a virtual orphan. Now confronted by the son, the hero, and contending for the heroine's affections, he plans another death. He does so despite constantly being overwhelmed by the horror of his crimes and, indeed, haunted by visions--'Memory, oh, memory of guilt and pain!'. Previously, a sense of contrition indicated that the villain was not beyond reformation, as in the case of Don Carlos in *Loretta*. This is not so with the Baron. His end is conventional (like the gipsy queen in *The Bohemian Girl*, he dies by the bullet he intended for his victim), but the internal conflict between remorse and evil shows the seeds of a new attitude.

In *The Lily of Killarney*, Hardress, an Irish landowner, is secretly married to the peasant-girl Eily O'Connor, but faces ruin unless he marries the heiress Anne Chute. His faithful servant, Danny Mann, misinterprets Hardress's wishes and attempts to kill Eily. In the act of drowning her, he himself is fatally wounded, and his death-bed confession incriminates Hardress in murder. Eily has in fact been rescued, and her dramatic appearance quashes the charge. Hardress acknowledges 'the Colleen Bawn' as his wife, and she is accepted by all--including Anne Chute.

Danny's character accommodates opposing emotions more forcefully than the Baron's. His aria owes its unique quality to a sophisticated mixture of tenderness for the heroine and loyalty to the hero: loyalty to the man who crippled him, and yet who exercises a fatal fascination. As he tells Eily:

> He made me a cripple--I bear him no ill--
> I lov'd him before, and I dote on him still.
> He might crush me to pieces, my last parting breath
> Would be to declare that I lov'd him till death.

Danny battles with his sense of remorse. He knows Eily's virtues: 'Heaven form'd in her a jewel rare.' But his love for Hardress and sense of duty over-rule morality:

> Can I sufficient courage find
> To harm a girl so gentle and so kind?
> ....
> Down cursed scruples! Hold thy peace, remorse!
> My duty to my master I'll fulfill
> Through good and ill
> Naught, naught can check me, well I know my course.
>
> Duty! Yes, I'll do my duty.
> What is love, and what is beauty,
> To a rough, misshapen creature,
> Crook'd in back, and hard in feature?
> Hearts that melt with soft compassion
> Beat in frames of other fashion.
> I'll help my master where I can;
> No other law has Danny Mann.

By squashing his conscience, Danny seals his fate. He comes closer to being a tragic figure than any other character in English opera of the period.

THE HERO

We shall need to devote less time to the hero, for the conventional conflict was played out by the heroine and the villain.

The essential features of the hero are identical to the heroine's: steadfastness in love, and reverence for virtue. Thaddeus, in *The Bohemian Girl*, for example, amply demonstrates heroic qualities of bravery and patriotism, but the crucial point is his nobility of spirit expressed in his love for Arline. This is in stark contrast to villains, who are bent on imposing their will on heroines. Arline thinks of him as the man 'who in every trial preserved [my] honor [*sic*], who twice restored me to a father's arms, and at length to a father's home.' He is the archetypal preserver of the society's values--chastity, family, home: to be welcomed by any Victorian parent as a prospective son-in-law.

One hero was based on a historical person: the Chevalier de St-Georges (1745-1799), portrayed in Balfe's *The Bondman* (1846) as the illegitimate son of an unscrupulous white man and a negro slave who achieves social nobility through his personal qualities. Doubtless this pleased anti-slavery movements, but the case was no different from so many others: Melésville and Roger de Beauvoir's 'comédie mêlée de chant' *Le Chevalier de Saint-Georges* (Paris, 1840) had been given in translation at the Princess's the previous year, and Bunn hoped to exploit the situation.

Like Max in *Der Freischütz*, a small number of heroes suffer lapses in their morality. Rudolph in *Lurline* and Hardress in *The Lily of Killarney* are the main examples. Hardress is the most interesting. His part was shaped by two factors: the character Boucicault inherited from Griffin, and the need to restrain his part in view of Harrison's role, the peasant Myles. Boucicault softened Hardress's outlines and ensured he survived to live happily with Eily, but the result is not entirely convincing: his stature is reduced, and his motivation is not explained with the same verve as marks Danny's part.

Most heroes are of noble birth. Nevertheless, like their beloved heroines, they prize virtue and love above wealth and power. This is such an important issue that they sometimes test heroines on the strength of their convictions. The following clear statement comes from *The Gaberlunzie Man*, written by William Rede and performed at an early stage in the period (1836) at Arnold's English Opera House.

> I have tried thee Lilian, and found thee all my heart ere [*sic*] hoped for, unswayed by name or pomp, undismayed by misery or danger. Oh, thou unbought gem, brightest, purest, greatest of Creation's work--Nature gathered all of her loveliness, and best, in one wondrous work, and called it woman.

In *The Rose of Castille*, hero and heroine test each other. Both are monarchs, and both have disguised themselves specifically for this purpose.

Thaddeus is unusual in that he is a political outcast. Bunn did not make much of the matter. Political issues are rare, and it arises here only because the librettist drew some features from a popular novel, Jane Porter's *Thaddeus of Warsaw*, which went through thirteen editions between 1803 and 1831. The English hero is neither a Mazeppa, ejected from a society for transgressing its mores, nor a Schubertian wanderer, looking back at the place of his emotional rejection. His goal is to gain the love of the heroine and with her establish a home--completely congruent with her wishes, and the fulfilment of the melodramatic ethos.

THE BALLAD

In many instances the hero voices his thoughts in a ballad. The ballad is so characteristic of English opera that it requires special attention.

The term has to be defined very expansively. Contemporary usage was too inconsistent to be a reliable guide. The word does not appear at all in the libretto of *The Bohemian Girl*. In the case of Lavenu's *Loretta*, one of the heroine's songs is listed as a 'Ballad' in the price-list of individual items printed with the libretto, but is called a 'Romance' in the libretto itself. To confuse matters still further, what the price-list calls a 'Romance' is a 'Grand Scena' in the libretto.

Put briefly, in musical terms, the ballad may be described as a simple melody in strophic form. It can be regarded as a sophisticated descendant of the folksong, retaining the latter's simplicity, though not its particular idioms. Indeed, the ballad took the place of the popular Scottish and Irish melodies frequently introduced in operatic works in previous periods. The most famous example from the earlier part of the nineteenth century, Bishop's 'Home, sweet home' from *Clari, or The Maid of Milan* (1823), both illustrates the relationship to folksong and yet is an exception to the rule: the composer claimed, at least initially, that it was derived from a Sicilian air, whereas ballads are normally newly-composed.

The musical simplicity and the textual characteristics of the English ballad distinguish it from the ballade of French opera, which has a more complicated musical style, a different poetic scheme, and strongly narrative element in the text, often based on a legend adumbrating the plot, as in Raimbaut's in Act I of Meyerbeer's *Robert le Diable* (1831) and others indebted to this type, such as Senta's ballade in *Der fliegende Holländer* (1843).

The ballad is associated with a certain mood and pace. It differs from other solo items in its generalized, reflective quality, and its slow and regular verbal rhythms. It may be introduced by speech or recitative, but is more often followed by speech, forming a separate unit that can easily be detached. The number of ballads per opera varies; in *The Bohemian Girl*, Thaddeus has two, Arline one, and her father one.

In broad cultural terms, the ballad represents a continuation of the eighteenth-century's equation of simplicity and virtue. (It will have been remarked in chapter 1 that simplicity was a quality admired in no less a person than the Queen of England.) The texts always express virtuous sentiments. Interviewed shortly before joining the staff of the Guildhall School of Music, Sims Reeves ('the greatest ballad-singer we have ever had') remarked that the ballad is 'but a sermon after all, and the singer is the preacher.' He prized simple qualities in a ballad, and implied that the appropriate vocal style was more taxing than opera and oratorio: 'If they are proficient at that they can tackle any ballad.' Balfe taught his singing pupils to 'study simplicity in *Ballads* or *Adagios*.' The librettist Samuel Lover maintained that simplicity was important in the text: emotion should be 'rather suggested than paraded.'[16]

One reason for its importance was commercial. There was an enormous domestic market for songs whose sales were promoted by public performances by the stars of the day. In 1840 Mark Lemon was paid £100 for his farce *Fashionable Arrivals* and an additional £30 for a song, set by Tully.[17] The pressure came directly from singers as well as publishers. Soane publicly complained in the libretto of *The Night Dancers* that he had been forced to include a song at a singer's insistence, despite its dramatic inappropriateness. He allowed it to be printed only in a footnote, explaining that 'The following song has been written for the sake of the singer [Bodda]; but, as it interferes with the interest of the scene, I have not introduced it into the text.'[18] By the mid 1860s, singers expected payment from authors for having popularized their songs.[19] The practice militated against close textual links between the ballad and the rest of the text, for specific references would make it less attractive to the general purchaser. At worst this led to bland anonymity, a glaringly detachable, dramatically irrelevant effect. A clear example occurs in the Lord Chamberlain's copy of Loder's *Francis I* (1838): a song which referred to the plot was replaced by another that omitted such specific reference.[20] A ballad for the tenor William Harrison was a staple ingredient of many operas. Sims Reeves, who succeeded Harrison as the leading tenor of the day, looked back at Balfe's *The Maid of Honor* with the following comment:

> Why ... was I made to sing a song about a chair, in no way connected with the opera, presumably introduced with the view of finding favour in the eyes of the music publishers? The sentimental song entitled, "In this old Chair my Father sat," proved, all the same, one of the successes of the work; indeed, its only success.[21]

There is a hypocritical note here. Lofty in his retirement, Reeves was biting the hand from which he had willingly fed.

The importance of the individual singer and of the commercial element is further emphasized by the case of Henry Phillips, the main bass in the early 1830s and early 1840s. The characteristic homilies of the ballad could only be appropriately expressed by the hero, heroine, or another virtuous character such as a father-figure, but there is a small number of ballads in villains' roles. Most were performed by Phillips. His part in Barnett's *The Mountain Sylph* (Hela) was originally conceived without a ballad, but he insisted on one: 'The ballad had by this time become a necessary appendage to all my representations, and a ballad I must have.' He agreed to a melody written some years previously by Barnett, words were quickly supplied, and, in complete conflict with the rest of Hela's character, 'Farewell to the mountain' was introduced in Act II. 'This was not the first instance in which a doubt existed in the mind of the management as to the necessity, or prudence, or policy, of a ballad being given to me in a similar position, nor the first instance of a song becoming popular which was not written for the work in which it was afterwards produced,' Phillips recalled. He jealously kept to himself 'The light of other days,' the ballad in Balfe's *The Maid of Artois* that Malibran coveted (he was 'too wise to part with so valuable a song as this'), and, as Charles Rice reported, later interpolated it into his performances of Caspar in *Der Freischütz*. He insisted on having a ballad in Bunn's production of Auber's *Gustavus III*. In this case the melody was selected from the overture, and introduced by a hastily-written recitative. Phillips then threatened to walk out if he could not sing it as he wished. On another occasion he was taken to court over the ownership of a song.[22]

As regards the ballad's technical features, Fitzball used a variety of poetic formats, but Bunn favoured a particular scheme which, thanks to his position, dominated in the 1840s and continued in use until the 1860s. This consisted of two verses, each of eight lines, with a hiatus at the end of the fourth. The form was not exclusive to Bunn, of course--Mark Lemon used it for 'In the days that are faded,' Phillips's ballad in *Fridolin*, one of the finest early examples--but, especially after the triumph of 'The Gipsy Girl's Dream' in *The Bohemian Girl*, Bunn's pre-eminence was openly accepted, to the point where Fitzball inserted two ballads by Bunn in *Maritana*, specifying they were written '*at the request of the Author and Composer*,' and two in his version of Scribe and St Georges' *Les Diamants de la Couronne* the following year (1846). Doubtless Bunn was happy to accept the pecuniary rewards; there may have been a long-term financial agreement.

Bunn found his model among the poems Thomas Moore wrote for his collections of national airs, thus reinforcing the general association between the ballad and the folksong. The compatibility of Moore's texts with musical structures was obvious, even though their own structures varied depending on whether they were of narrative or reflective mood. A particular pattern consists of two eight-line verses, frequently with the opening lines end-stopped (they are often in the rhythm -/-/-/-/ | -/-/-/), and, sometimes, with similar final lines. Moore also used this pattern for sacred texts, further contributing to the religious ethos Reeves later noted. On occasion there is a poignant air of sadness, of melancholy memory.

Bunn exploited all these features and thereby stamped his mark on the operatic ballad. The repeated final lines produce some similarity to the couplets of a romance in a French opera, but Moore's influence is uppermost. The use of Moore's opening rhythms is clear in Thaddeus's 'When other lips, and other hearts,/Their tales of love shall tell,' and Alberto's 'At morn upon the beach I stood,/And saw the waves depart' in *The Brides of Venice* (set by Benedict, and produced in 1844). Arline's 'Dream' adds extra unemphasized syllables: -/- - /-/-/ | -/- - /- - /. The generalized preoccupations of Moore's verses were exactly as required, though Bunn was usually careful to achieve a balance between the general and the specific, the contextually suitable and the commercially functional. There are no specific names or detailed references, but there is a restrained reaction to a moment of heightened emotion, summarizing the individual's feelings and casting them in a broader, generalized perspective. This produces a sense of emotional distance, allowing the character to come across as controlled and dignified, with a reflective quality born of sober experience. Often there is a progression that is sometimes found in Moore: personal sadness (verse 1) prompting generalized, aphoristic comment on adversity, and final maxim, often in the verbal formula 'Though x ..., y' (verse 2). Alberto's ballads in *The Brides of Venice* illustrate this.

Verse 2 of his ballad 'If a tear should repose' begins 'Though the world should forsake'; in 'At morn upon the beach I stood' the words are 'And thus it is with life--its cares/Are like yon mighty sea.' This is Reeves's homilistic element.

Bunn and Fitzball had distinct preferences as regards the imagery of their ballads. Fitzball concentrated on the 'humble cot,' making a speciality of picturesque detail and projecting a comforting background of English family and village scenes. Lyonnel's ballad 'Behold the happy home, where my forefathers dwelt' in *The Maid of Honor* is an example. The Rousseauesque values are overladen with images of family stability, village church, and loved ones long dead--images so potent (especially to the city-dweller) that they were imitated by Loder the following year in *Robin Goodfellow* ('My village church where first I heard').

The heroine's role gives fewer occasions to voice such sentiments. Lacking the experience acquired by the hero, she does not expound homilies so readily: this is almost a male prerogative. Her ballads emphasize her sense of values and her love for the hero. The effect of the characteristic pace and reflective element of the ballad are no less noticeable, as is made clear by a comparison of the French and English versions of Elvira's part in Adam's *Le Muletier de Tolède* (1854) and Balfe's *The Rose of Castille* (1857). When Harris and Falconer came to d'Ennery and Clairville's chanson (Act II sc.vi) beginning

Au couvent
Bien souvent,
J'ai pu rire
Maintenant
Je soupire,
En rêvant,
Au couvent.

they replaced it with 'The Convent Cell'--whose first verse is as follows:

Of girlhood's happy days I dream,
My home the house of prayer,
As in the bosom of a stream
Seemed heaven reflected there.
In regal halls, where oft I sigh,
Fond memories with me dwell

Of many a blissful hour gone by,
Pass'd in my convent cell.

The French version presents a bright, animated, young woman stifled by courtly etiquette, its text sparkling with fast rhythms and wittily unpredictable repetitions. The English version is quite the opposite, leading to inconsistency of characterization, for such seriousness is not expressed anywhere else in Elvira's role.

The most famous of all ballads is 'I dreamt that I dwelt in marble halls' in *The Bohemian Girl*. Its popularity was immense. Dickens's correspondence contains frequent references to it, and sixty years later James Joyce could have a character confuse the order of the verses, confident that his readers would not need a detailed explanation ('Clay,' in *Dubliners*). It reads as follows:

I dreamt that I dwelt in marble halls,
With vassals and serfs at my side,
And of all who assembled within those walls
That I was the hope and the pride.
I had riches too great to count--could boast
Of a high ancestral name;
But I also dreamt, which pleas'd me most,
That you loved me still the same.

I dreamt that suitors sought my hand;
That knights, upon bended knee,
And with vows no maiden heart could withstand,
They pledged their faith to me.
And I dreamt that one of that noble host
Came forth my hand to claim;
But I also dreamt, which charm'd me most,
That you loved me still the same.[23]

On one level it is a conventional affirmation of values and love. But its play on differing states of consciousness raises it above the norm. Arline's present and past are intertwined, and the only realities are the marble halls and Thaddeus--symbols of her childhood and her adulthood. These are brought together by a dream. Dreams were of much contemporary interest, and for an author they had obvious structural advantages, exploited in broader terms in Loder's *The Night Dancers*, Mellon's *Victorine* (1859), and, in a different medium, Dickens's *The Christmas Carol*, which appeared in the same year as

The Bohemian Girl. The vocabulary and imagery were not unprecedented. 'Halls' and 'walls' are monosyllables that rhyme conveniently, as Moore's 'The harp that once through Tara's halls' (which Bunn quoted in his autobiography)[24] illustrates; and De Luna's aria in Rooke's *Henrique* (1839), 'Methinks I tread my father's halls/With childhood's happy step and grace ...,' has similar images to 'The Dream.' But Bunn's particular merging of past and present and the dialogue with a second person, a combination that makes 'The Dream' unique among English ballads, brings to mind another case: Mignon's 'Kennst du das Land' from Goethe's *Wilhelm Meisters Lehrjahre*. In Carlyle's translation, the second verse has similar images and two dialogues--one with the occupants of the halls, another with the guardian: 'Know'st thou the house, with its turretted walls,/Where the chambers are glancing, and vast are the halls?/Where the figures of marble look on me so mild ....' It is impossible to make categorical assertions, but it is pleasant to conjecture that the most famous English song of the nineteenth century derives from one of the most famous German song texts.

NOTES

1. *The Melodramatic Imagination* (New York, 1985), p.xiii.
2. See Charles Rice, *The London Theatre in the Eighteen-Thirties*, edited by Arthur Colby Sprague and Bertram Shuttleworth (London, 1950), p.38.
3. *Rambler*, Essay no.4, 31 March 1750, p.24.
4. *Essais sur la peinture* (Paris, 1795 [1765]), p.78.
5. *Athenaeum*, no.680, 7 November 1840, p.887.
6. In this connection, it is interesting to note that in his translation of *Rigoletto* Oxenford rendered Gilda's 'Ciel! dammi coraggio!,' prior to her account 'Tutte le feste al tempio,' as 'Oh! *shame* [my emphasis] give me courage,' whereas 'heav'n' (chosen by Natalia Macfarren in her translation) is more accurate and as easy to sing.

7. *The Children of the Abbey*, 4 vols (London, 1796), 2:255.

8. Johann Lavater, *Essays on Physiognomy*, translated by Thomas Holcroft, 18th ed. (London, [1885?]), p.99.

9. Chapter 51, 'A Charade is Acted,' *passim*.

10. Henry Scott Holland and W.S.Rockstro, *Memoir of Madame Jenny Lind-Goldschmidt*, 2 vols (London, 1891), 1:[3]. Doubtless Holland's position as Canon and Precentor of St. Paul's Cathedral gave their views all the more weight.

11. *Victorian Prose and Poetry*, edited by Lionel Trilling and Harold Bloom (New York, 1973), p.331.

12. Henry Home, Lord Kames, *Elements of Criticism*, 3 vols (Edinburgh, 1762), 1:62 fn. This work continued to be popular well into the nineteenth century, reaching its eleventh edition in 1839.

13. *Sketches by Boz*, Gadshill Edition, 2 vols (London, 1898), 2:241-42.

14. Quoted from the 'second edition, corrected' of the *Songs, Duets, Choruses, &c. in The Mountain Sylph* [London, 1834].

15. I. 1. Cf. *Marie Tudor*, Journée I, sc.vii.

16. Sims Reeves: undated newspaper cutting [1888], BL, Add.35027, f.111v.; Balfe, *Ballad Tutor, without the use of Solfeggi* (London, 1883), p.[i]; Bayle Bernard, *The Life of Samuel Lover R.H.A.*, 2 vols (London, 1874), 1:148.

17. Arthur A.Adrian, *Mark Lemon: First Editor of Punch* (London, 1966), p.25.

18. *The Night Dancers* (London, 1846), p.19.

19. See Bernard, 2:181-82.

20. BL, Add.42949, ff.391v-92.

21. Sims Reeves, *My Jubilee, or, Fifty Years of Artistic Life* (London, [1889]), p.77.

22. Henry Phillips, *Musical and Personal Recollections*, 2 vols (London, 1864), 1:192-94, 217, 240-42, 201-12; Rice, p.61.

23. Quoted from the earliest vocal score, published by Chappell [1843?]. I have supplied the punctuation of verse 1, lines 4 and 6, and verse 2, lines 4 and 8. The capitalization and punctuation and the spelling of 'loved' and 'pledged' were altered ('lov'd' and 'pledg'd') in the vocal score edited by Arthur Sullivan and J.Pittman and published by Boosey (London, [1872]).

24. Bunn, 1:xxii.

III

The Librettists' Craft

The initial task facing a librettist was the selection of a plot. Most English libretti of the period were based on foreign works, particularly French operas and ballets. British works were not often used. The days of Scott's novels being 'Terryfied' for the stage, for instance, had passed (Daniel Terry had adapted *Guy Mannering* for Bishop in 1816), and *The Lord of the Isles* by Rodwell and Fitzball (1834) and *Quentin Durward* (Laurent and Fitzball, 1848) were rare cases. However, *Ivanhoe*, whose plot is entirely melodramatic and which had the additional benefit of two contrasted heroines, was the model for a similar arrangement in Benedict's *The Gipsy's Warning* (1838) and *The Crusaders* (1846), whose very title is indicative. Two later operatic versions of Scott's tales were Balfe's *The Knight of the Leopard*, performed posthumously in 1874 (based on *The Talisman*) and Sullivan's *Ivanhoe*, of 1891.

The melodramatic genre could make little use of Byron's tragedies. Romer's *The Pacha's Bridal* (1836) has clear overtones of the *Corsair*, but the case of *The Siege of Corinth* (1836) is symptomatic. As Planché explained in the libretto, this was less a direct adaptation of Byron's work than an arrangement by Cooke of Rossini's *Le Siège de Corinthe*, the Paris version of *Maometto II*. Other native authors were used equally infrequently. There is an obvious Shakespearian background to Loder's *Robin Goodfellow, or, The Frolics of Puck*, for which he wrote his own libretto (1848). However, the *Midsummer Night's Dream* element is heavily diluted. Titania orders Puck to discover what is dearest to a mortal woman's heart (he eventually discovers it is her will), but this runs parallel with a Restoration plot in which a Royalist contends with an evil Puritan for the heroine's hand. The only English operatic setting of Shakespeare in this period was Balfe's *Falstaff*, performed, significantly, not at an English theatre but at Her Majesty's, in Italian (1838). The libretto was by Maggioni, the house translator.

Fitzball drew on Southey's poem for Balfe's *Joan of Arc*, but this also owes much to the melodrama of the same title that he had produced in 1822. The work was a pawn in the cut-throat rivalry between Bunn and Macready, who had just become manager of Covent Garden. Serle's play *Joan of Arc* beat Balfe's opera by two days (28 and 30 November 1837), a battle duly satirized by Planché in his revue *The Drama's Levee, or, a Peep at the Past* (1838), in which the two Joans sing an antagonistic duet to the strains of *The Beggar's Opera*.

Adaptations of gothic novels were also no longer frequent, though elements of the genre had continuing influence. Two operas were based on 'Monk' Lewis's works: Loder's *Raymond and Agnes*, which Fitzball derived from an episode in *The Monk*, and Balfe's *Bianca, the Bravo's Bride*, based by Simpson on *Rugantino*. (A ballet version of *Raymond and Agnes* had been given in 1797, and Lewis's own stage version was produced in London in 1811.) A later type of novel, the Newgate tale, barely made an impression. Fitzball's adaptation for Rodwell of Bulwer's *Paul Clifford* exploited the novel's immediate popularity, but swiftly sank into obscurity. As part of the rivalry between the theatres, it opened at Covent Garden on 28 October 1835. The next night saw the premiere of Balfe's *The Siege of Rochelle*, of which Fitzball was also the librettist, at Drury Lane.

The operatic use of oriental tales also waned during the early years of the period. Arnold adapted Loder's *Nourjahad* (1834) from the text he had written in 1813 for Kelly (*Illusion; or the Trances of Nourjahad*), itself based on a Persian tale by Mrs Sheridan. Mary Russell Mitford, author of Packer's *Sadak and Kalasrade* (1835), took the public's familiarity with the story of Sadak for granted ('who is there so dull or so wise as not to be familiar with that gorgeous Eastern fiction?' she asked rhetorically in the preface), but Hullah's *The Barbers of Bassora* (1837) was probably the last example. (The oriental setting made a late reappearance in Macfarren's *The Sleeper Awakened*, of 1850, but this is a 'serenata' consisting of four scenes only.) Another enthusiasm that weakened during the period was for the supernatural, which faded sooner in opera than in the novel. The demonic element became associated with Germany, the land of *Der Freischütz*.[1] Female spirits, primarily the figures of Giselle and Undine, stemmed from the world of ballet, focusing on the theme of the love of mortal man and supernatural woman. This underlies

Barnett's *The Mountain Sylph* (1834), Balfe's *Keolanthe* (1841) and *Satanella* (1858), Loder's *The Night Dancers* (1846), and Wallace's *Lurline* (1860). No copies of the libretto of *Keolanthe* are known. Balfe's biographer, Kenney, gives a brief account of the plot, which involves a young student from Padua and a princess from ancient Egypt (Keolanthe).[2]

Operas set in the British Isles were small in number, but they include some of the most significant works of the period, particularly *The Mountain Sylph*, set in Scotland, and Benedict's *The Lily of Killarney*, the scene of which is laid in Ireland. Macfarren took a special interest in English settings. His early works, *The Devil's Opera* (1838) and *An Adventure of Don Quixote* (1846), take place abroad, but his remaining six operas, from *King Charles II* (1849) to *Helvellyn* (1864), are set in England, including the only adaptation of Oliver Goldsmith, *She Stoops to Conquer* (1864). Three of Balfe's operas, *Catherine Grey* (1837), *The Maid of Honor* (1847), and *The Puritan's Daughter* (1861), have an English background. The appearance of 'madrigals' in several operas gives a national flavour, but is best considered a continuation of the glee of Bishop's time.

With the exception of the latter point, the changes in taste were due not only to natural fluctuations in popularity but also to an increasing reliance on French works. In the copy he submitted to the Examiner of Plays, Rodwell described his *The Seven Maids of Munich* (1846) as 'an Original Musical Afterpiece in One Act and not from the French,' and though he later omitted the jibe, it was well founded. ('Musical Afterpiece'--a subsidiary item in an evening's entertainment, standard since the eighteenth century--was altered to 'Operetta,' clarifying the latter term's meaning.) The extensive use of French sources was primarily due to Alfred Bunn and his dominating position at Drury Lane. Balfe's works reflect this very clearly. At least eleven of his English operas were based on a French opera, ballet, play, or novel.[3]

Bunn's admiration for French works was understandable. If there was a single capital of opera, it was Paris. The same could be said regarding ballet. Much of this reputation was due to the Parisian librettists, led by Scribe and St Georges. They were masters of grand opera, opéra comique, and ballet; craftsmen with enviable skills of construction and command of the mot juste, sought by many a composer and impresario. Wagner reported that fifty-two German impresarios attended the first night of Halévy's *La Reine de Chypre* at the Paris Opéra in 1841.[4] As noted in chapter 2, the libretto was by St

Georges. Doubtless Bunn, or an emissary such as Planché, was also there. One result was that *La Reine de Chypre* appeared in four other versions in different parts of Europe within five years, with music composed by Lachner (1841), Donizetti, Balfe (both 1844), and Pacini (1846). Another example is provided by Flotow's *Martha* and Balfe's *The Maid of Honor*, both of 1847, each derived from St Georges's and Mazillier's ballet *Lady Henriette* (1844)--a collaborative work for which Flotow himself had written the first Act.

There were recurrent complaints that English writers suffered as a result of this reliance on foreign works. George Soane, who, it will be remembered, was the librettist of Loder's *The Night Dancers*, spoke of the 'many other foreign productions, with which the English have been surfeited, to the almost total exclusion of native talent.' But this was in the preface to his own translation of a French work, Scribe's *Haydée* (set by Auber), which, he admitted, was written 'with all the skill of a master.' The easy access to such plots was an overwhelming temptation. Translations of proven Parisian successes were quicker and easier for the author and safer for the manager.[5] Planché pointed out that the English stage had long been indebted to some degree to the French, and deeply resented the lowly status of the translator. He attributed the need for translations to the incessant craving for novelty.[6] Fitzball insisted that it was important for the audience to be familiar with the plot of an opera, and so the existence of an earlier version already known in London was beneficial. 'Few subjects become so popular, musically adapted to the stage,' he wrote in his 'Advertisement' to Wallace's *Maritana*, 'as those with which the Public are already familiar.' *Maritana* was a good example: as noted previously, it was based on d'Ennery and Dumanoir's *Don César de Bazan*, itself loosely derived from Hugo's *Ruy Blas*.[7] *The Night Dancers* is another instance. Soane had considered adapting the German myth of Giselle for an opera some twenty years earlier, but abandoned the idea because 'in those days there was no native composer able to carry out my views.' (Bishop, who had set Soane's *Aladdin* in 1826, must have been mortified to read this in the libretto). He revived the idea after performances of the French ballet *Giselle* at the Princess's, the theatre where *The Night Dancers* was produced.

We may now turn in more detail to the procedures followed by librettists.

All libretti were submitted to the Examiner of Plays, appointed by the Lord Chamberlain. English libretti caused little official complaint.

Aware that Auber's *La muette de Portici* had incited the revolutionaries in Brussels in 1830, William IV was advised to cancel a royal command performance of the opera,[8] but this was an isolated case; the English theatre was recognized as being apolitical. Often the copies delivered to the Lord Chamberlain's office were dated, and it is clear that they were frequently submitted only a few days in advance of the first performance and then quickly returned. On occasion there was doubtless a measure of self-restraint. It would not have been in the interests of the management to submit texts knowing that they might be rejected. For example, historical details of the French king's adultery were avoided by Mary à Beckett and her husband Gilbert in *Agnes Sorel* in 1835, but in 1837 John and Charles Barnett (not related) were unequivocal about the adulterous King Henry II in *Fair Rosamond*, which includes an interrupted coronation scene in Westminster Abbey and a queen who attempts to kill her rival, the king's mistress. Though this was a reworking of a plot written by Addison (*Rosamond*), first set by Clayton in 1707 and subsequently by Arne and Arnold, managements may have considered it prudent to wait before proposing the subject. The important factor was the replacement of George Colman by the more liberal Charles Kemble as Examiner of Plays in 1836. Colman had been notorious for his restrictive outlook where mores were concerned. Equally, it would have been unwise to produce operas that might antagonize the audience, especially in an age of clear-cut views on the morality of the stage and when religious feeling was very strong. A rare case when Planché misjudged, and *The Times* acted as censor, will be noted shortly.

Gauging from Fitzball's own copy of *Diadesté*, which he submitted to the Examiner of Plays, his practice was to send the libretto to the composer (in this case Balfe) in sections, as the work progressed. His method of dramatizing a novel was to draw up a 'sketch' while his wife read it aloud. From the novel he would get 'the essence of what I required, and I could draw upon my own imagination for the rest.'[9] He spoke about his technique of adapting a foreign opera during the hearing of the case Bunn brought against Jenny Lind for breaking her contract to sing at Drury Lane. He took five weeks to produce a translation of Meyerbeer's *Ein Feldlager in Schlesien*, for which Bunn paid him 'about £150.'[10] He was told to adhere as closely as possible to the incidents, but he might alter the language. He had a French and a German libretto at hand. A German dictated a literal translation, and Fitzball 're-organized' it into verse and prose. He had to write a new ending, for in the original 'Mademoiselle Lind was made to stand up with a wand, and describe a phantasmagoria on the back of the stage, representing the burning of a theatre, in which the English took no interest.' (*Ein Feldlager* was performed on the re-opening of the Berlin opera house in 1844, after it had been destroyed by fire.)

The adaptation of a ballet required a different approach, for the crowd scenes and the locations of the main individual displays already existed. Bunn's adaptation of St Georges's and Mazillier's *La Gypsy* (1839) to form *The Bohemian Girl* illustrates his methods. The ballet was based on Cervantes's novel *La Gitanilla*, which had already served as the basis for Weber's *Preciosa* (1821). With Scott's novels in mind, St Georges altered the setting and period from Renaissance Spain to Restoration Scotland, and divided Cervantes's continuous narrative into two sections--the kidnap of the young heroine, and, later, her life in the gipsy community, her trial (in Edinburgh), the gipsy queen's jealousy, and the denouement. St Georges also gave the hero the character of a proscribed soldier.

Bunn changed the location again. His choice of Austria and of a Polish hero was derived from Jane Porter's *Thaddeus of Warsaw*. Set in the 1790s, the novel tells the tale of a brave young Polish count who has to flee his beloved Poland, now assailed by Austria and Russia. It is replete with secret marriages (including a trip to Gretna Green--much of the story is set in Britain) and the conflict of virtue versus evil, with the eventual downfall of the latter. Bunn exploited two sources of success: the strength of a Parisian plot, and the familiarity of a popular novel.

An obvious task was to replace displays of individual dancing with individual singing. Bunn also added the recurrent gipsy chorus ('In the gipsy's life you may read,/The life that all would like to lead'), and altered other details, such as delaying the hero's appearance in the trial scene so that all attention is focused on the heroine and her judge, in fact her unknown father. The ballet's synopsis included occasional brief explanatory speeches, and Bunn retained these, with some emendations. Contriving the absence of the hero so that he could not assist the heroine at a crucial time was a common task. Several operas used the device of imprisoning him on a false charge.

The choice of which sections of the text should be set to music, and which spoken, depended on several factors--the work on which the libretto was based, the composer, manager, and the performers. In *The Bohemian Girl*, by no means Balfe's most extensive opera, Act I has only three sections of speech, Act II six, and Act III three; several of these consist merely of a few lines. The spoken sections have various functions. In the early stages of the work they are used to establish characters' identities and roles. The developments of the plot that are delivered in speech are Thaddeus's rescue of Arline (as a child) in Act I, and the accusation of Arline for theft in Act II. After an opening dialogue, Act III is virtually continuous music.

As a comparison of Boucicault's *The Colleen Bawn* and *The Lily of Killarney* shows, the adaptation of a play required the reduction of spoken dialogue and the reordering of material that characterizes an individual. A play may spread over a broad canvas various lines that build the impression of a personality, but in an opera these need to be collected into larger, discrete units, especially arias. This draws attention to an issue that all librettists needed to consider, regardless of the type of work to be adapted: the requirements of the singers.

The importance of this element cannot be underestimated. Danny Mann's aria in *The Lily of Killarney*, written specifically to suit Charles Santley, has already been mentioned. It is among the grandest of all the arias in English opera of the period. Singers and composers made various demands. Towards the end of his career, William Harrison's voice declined to the point where he could no longer sustain the hero's part; operas produced under his joint management, therefore, had to have a role written with this in mind. In *The Lily of Killarney* he took the part of Miles, the warm-hearted Irish peasant, and in Balfe's *The Puritan's Daughter* he played a friend of King Charles II--neither being the hero, who was played by Santley. Even so, and like many other singers, Harrison demanded a ballad, another requirement to which the librettist had to respond.

Recitative presented a particular task. English recitative differed from its Italian and French counterparts in terms of its poetic character. The standard pattern for Italian recitative was versi sciolti, a mixture of seven- and eleven-syllable lines that allowed a degree of almost conversational flexibility, shared, to some extent, by the French vers libre, with its Alexandrine background. By contrast, English recitative was written in heroic couplets, a usage that extends back beyond Arne, and has all the formality of the poetic idiom of Pope and Dryden. The pronounced stresses and rhymes give rise to a rigidity and recurrence of emphases that set the national recitative apart from its continental equivalents. This is particularly noticeable when, as in the case of Bunn's couplets, the literary quality is low, as was exemplified in chapter 2. A higher standard was achieved by Bridgeman in Balfe's *The Armourer of Nantes*. The stilted effect caused by Bunn's vocabulary is largely absent, the sense flows more easily from line to line, and the movement from one character to another is smoother.

Matters concerning the ballad, aria forms, and so forth, are considered elsewhere in this study, and here it will be better to turn to another aspect of the conventional libretto: the ending. The denouement had to ensure the required triumph of virtue at the last possible juncture. Two attempts to disregard the norm indicate the strength of the convention. Planché was eternally aggrieved by the treatment of *The Red Mask*, his adaptation of Marliani's *Il bravo*, produced at Drury Lane under Bunn's management in the same year as its Parisian premiere, 1834. Fenimore Cooper's novel *The Bravo* (1831), on which it was based, ends with the decapitation of the hero. The two previous London versions (melodramas) had reprieved him at the very last moment, but Berretoni, Marliani's librettist, had followed the novel's ending, and so, initially, did Planché. The first performance concluded with the crowd separating to reveal the executioner standing beside the block, with the victim's body covered by a black cloth. There was uproar. A sub-editor of *The Times* immediately told Bunn that he must alter it. This was on a Saturday night. On the following Monday morning *The Times* carried a long review, detailing the 'shocking mistake,' and informing its readers that the ending had been changed over the weekend. A year later, Planché's version of Halévy's *La Juive* was given at Drury Lane (alongside Balfe's *The Siege of Rochelle*), and although he altered this, too, he explained the original conclusion and announced his regret 'that truth, power, and poetical justice should be all sacrificed, as in the recent case of the "Red Mask," to a prejudice.'[11]

Ten years after *The Red Mask*, in 1844, Bunn publically bowed to this English prejudice in the libretto of *The Daughter of St Mark*, explaining in the preface that 'I was apprehensive of introducing any representation of those fearful events which signalized *Cyprus* in 1473, and which, while forcefully delineated by the pen of this able writer [St Georges] and through that medium too faithfully imitated on the Parisian stage, are repugnant to the taste of an English audience.' In fact, the most surprising instance of an ending altered to conform to convention is in his own *The Bohemian Girl* of the previous year. Cervantes's tale ends with the marriage of hero and heroine, her false accuser being forgiven. St Georges's ballet ends as follows: Lord

Campbell happily accepts Stenio as Sarah's future husband, leaving Mab (the gipsy queen) mad with rage. She orders a gipsy to shoot Stenio--and he falls dead. In Sarah's eyes, Mab becomes a supernatural, hellish Fury, repeating her fateful prophecy '*qu'il ne serait jamais à toi*!'. With all the passion of a gipsy, Sarah throws herself at Mab and stabs her, and then runs to Stenio--and falls lifeless in her father's arms. The dying Mab looks on scornfully.

Bunn's original intention was to retain the unhappy ending. The libretto submitted to the Examiner of Plays ends with the gipsy marksman killing Thaddeus, whereupon the queen stands over his body and repeats her earlier threat to Arline: 'Weep rivers--for ages pine/He shall never be thine.' The curtain falls after the final appearance of the gipsy chorus--an extraordinary idea at this juncture. Exactly when, or after what discussions, the ending was changed so that the queen, not Thaddeus, is killed, we cannot know, though the alteration must have been made after 15 November 1843, when the Examiner received his copy.[12] (The first performance took place on 27 November.) The revision suffers from the abrupt alteration: there is no comment on the event, and the rondo finale that Bunn now had to write for Arline has only the standard expressions of joy, regardless of the death that has just occurred. Nevertheless Miss Rainforth, who performed the part, doubtless strongly preferred having the conventional display-piece, and her wishes must have been influential in the decision to alter the ending. The revised version was also used in Germany, but the ending was again changed for performances in Trieste (1854) and Rouen and Paris (1862 and 1868). In the Italian version the queen survives, but is arrested. In the French version she dies, overcome with remorse, the object of universal sympathy.

NOTES

1. See the 'Remarks' in C.A.Somerset's version of Weber's *Sylvana*, in *Cumberland's Minor Theatre*, 16 vols (London, 1828-44), 3:4.

2. Kenney, pp.150-51.

3. *The Siege of Rochelle*, *The Maid of Artois*, *Diadesté*, *The Bohemian Girl*, *The Daughter of St Mark*, *The Bondman*, *The Maid of Honor*, *The Rose of Castille*, *Satanella*, *The Armourer of Nantes*, and *Blanche de Nevers*. The number may be higher--for instance, *Keolanthe*'s plot could well have been based on a ballet, and there may have been a French element in *The Enchantress*.

4. *Wagner Writes from Paris*, edited and translated by Robert L.Jacobs and Geoffrey Skelton (London, 1973), p.164.

5. See the 'Remarks' prefacing Mrs Frances Davidson's translation of Scribe's *Giralda* (London, [1850], pp.[5]-7.

6. Planché, 1:246.

7. It is interesting to note that the colourful character of Don César also attracted Verdi, rather than Ruy Blas himself. See Franco Abbiati, *Giuseppe Verdi*, 4 vols (Milan, 1959), 4:380.

8. *Report from The Select Committee appointed to inquire into the Laws affecting Dramatic Literature* [London, 1832], q.3946, p.219.

9. *Thirty-Five Years of a Dramatic Author's Life*, 2 vols (London, 1859), 1:154-55, 234.

10. Alfred Bunn, *The Case of Bunn versus Lind ... February 22nd, 1848* ... (London, 1848), p.38.

11. Libretto of *The Jewess* (London, 1835), p.48. See also the *The Times*, 17 November 1834, p.3; Planché, 1:218-22, 243-44.

12. See BL, Add.42970, f.457v. A note on f.427 indicates that the Examiner's copy was received on 15 November and dispatched on 16 November.

IV

The Operatic Background

The purpose of this chapter is to give an account of the historical background of English opera insofar as it affects the period in question. It will conclude with some preliminary thoughts on the musical styles used in the period, in order to explain the approach to be adopted in the more detailed study that will follow.

THE ATTITUDE TO ITALIAN OPERA AND THE DEVELOPMENT OF THE ENGLISH ALTERNATIVE

Although strongly supported in various quarters, particularly among the aristocracy, Italian opera had been the target for sustained attack from the early years of the eighteenth century. John Dennis's *An Essay on the Opera's After the Italian Manner* of 1706 may have been extreme ('an Opera after the Italian Manner is monstrous ... so prodigiously unnatural, that it could take its beginning from no Country, but that which is renowned throughout the World, for preferring monstrous abominable Pleasures to those which are according to Nature'[1]), but his dislike was supported by Addison in *The Spectator* in 1711 and Pope in *The Dunciad* of 1742 ('When lo! a Harlot form soft-sliding by').[2] Opposite views were, of course, expressed--Burney springs to mind--but the censure continued into the nineteenth century. William Hazlitt, a frequent contributor to *The Examiner* and similar journals in the early part of the century, gave voice to biting criticism of Italian opera in two essays, 'Sir Walter Scott, Racine, and Shakespear,' in which he raised the question 'Whether we receive more pleasure from an Opera or a Tragedy,'[3] and 'On the Opera.' He poured scorn on the mixture of aural and visual media, the minimal imitation of nature, and the way in which music impeded the object of theatrical presentation--to evoke terror and pity. The influence of beliefs drawn from Aristotle, Addison, and Schlegel is evident. An important plank in his argument was the conviction that poetry was superior to music. This hallowed belief was shared by many in a country that could look with pride on its literary heritage, and, naturally, was upheld all the more fervently by writers--the most articulate group in society. Hazlitt voiced his opinion very forcefully. Italian opera, he wrote,

> is a species of intellectual prostitution; for we can no more receive pleasure from all our faculties at once than we can be in love with a number of mistresses at the same time.... It does not subsist as an imitation of nature, but in contempt of it; and instead of seconding, its object is to pervert and sophisticate all our natural impressions of things ... every appeal to the imagination is sheathed and softened in the melting medium of Siren sounds. The ear is cloyed and glutted with warbled ecstasies or agonies; while every avenue to terror or pity is carefully stopped up and guarded by song and recitative. Music is not made the vehicle of poetry, but poetry of music ...[4]

Seen against the background of this deep-seated attitude towards Italian opera and the role it allocated to music, the development and continuing support for an alternative form using English texts, partially delivered in speech, was predictable. The tradition extended to the seventeenth century; with the exception of *Dido and Aeneas*, all Henry Purcell's theatrical works include spoken material. (In *King Arthur*, for instance, none of the principal characters sings: the drama is not developed in the music, which is incidental to the plot.) Similar developments occurred in other countries, incorporating other forms of popular entertainment and taking place in theatres of more modest financial means, attended by the bourgeoisie and lower classes. The opéra comique, the singspiel, and the zarzuela are examples. In the early eighteenth century, the division between the Italian and English forms of opera became extremely clear, with Handel's opere serie on the one hand and the ballad opera on the other. In the second half of the century, the repertory of English opera was developed by a number of composers--Arne, Arnold (father of the theatrical manager), Dibdin,

Shield, Storace, and others--often with pasticcio works, such as *Love in a Village* (1762), with music by Arne and sixteen others, and *The Duenna* of 1775, a compilation of items by nine composers, including two members of the Linley family (Thomas Linley, father and son).

Any attempt to put English opera on the same footing as the Italian model by setting all the dialogue to recitative encountered stiff opposition. In *Some Reflexions Concerning Operas, Lyric Poetry, Music, &c.*, accompanying the libretto of *Rosalinda* (set by J.C.Smith in 1740), the author, John Lockman, conceded that 'I am not ignorant, that one great Objection is frequently made to the Manner in which this Part of a musical Poem is delivered, as being quite unnatural. It must be so on many Occasions.'[5] His only defence was that a form of recitative was used by the Greeks and Romans. Of the operas that were through-composed, Arne's *Artaxerxes*, a translation from Metastasio (1762), was the most widely known. Arne's comment in the preface on recitative is restricted to Metastasio's poetical style. Some of Arnold's operatic works were all-sung, the last being *The Enraged Musician* of 1789 (he continued to write for the stage until 1801), but they were short works, such as afterpieces. General Burgoyne, the librettist of William Jackson's 'comic opera' *The Lord of the Manor* of 1780 (loosely based on Marmontel), and, significantly, also a playwright, was not only scathing in the preface[6] of his libretto about Italian and French serious opera (Metastasio ignored the 'laws of regular drama'; French opera was 'if possible even more absurd than the Italian in its departure from probability') but also doubted that recitative in English could ever be successful. English had a 'limited expressive range of normal speech' compared to Italian, and,

> in proportion as the ordinary expression comes nearer monotony, recitative or musical dialogue, will seem the more preposterous.
>
> .... If any specimens can yet be produced of it's having been effected, they will be found to consist only of a few lines introductive of the air which is to follow, and as such received by the ear just as [a] symphony[7] would be. Very few serious pieces, except Artaxerxes, can be recollected upon our Theatre where it has not entirely failed, even when assisted by action; in Oratorios it is, with a few exceptions, and those sustained by accompaniment, a soporific that even the thunder of Handel's chorusses are hardly loud enough to overcome.

Emphasizing the importance of holding 'a mirror up to nature,' Burgoyne also insisted that no dramatic action could be set to vocal music:

> vocal music should be confined to express the feelings of the passions, but never to express the exercise of them.... Music, therefore, if employed to express action, must be confined to dumb shew ... care must be also taken, that it does not interrupt or delay events for the issue of which the mind is become eager. It should always be the *accessory* and not the *principal* subject of the drama ...

This view is, of course, the antithesis of the principle underlying Mozart's ensemble finales. To see it in perspective, we may recall that *Le Nozze di Figaro* was first performed only five years after Burgoyne's preface and libretto were printed.

In all likelihood, there was no all-sung English opera between the pasticcio *Dido, Queen of Carthage* of 1792, comprising music by Storace and others, and C.E.Horn's *Dirce, or, The Fatal Urn*, based on Metastasio's *Demofoonte* (1821). The next was Balfe's *Catherine Grey*, produced in 1837.[8] After his *The Daughter of St Mark* (1844), no further attempts were made for over thirty years, until Sullivan's *Trial by Jury* (1875). For works on a larger scale we have to turn to the 1880s--Goring Thomas's *Esmeralda* and Mackenzie's *Colomba* (both 1883), and Stanford's *The Canterbury Pilgrims* (1884). Again, the management (in this case, the Carl Rosa Company) was a significant factor. An important manager from the earlier part of the century, Arnold, never felt able to commit himself to opera that was entirely set to music. In 1825 he wrote to the enthusiastic young Gerald Griffin (author of the novel *The Collegians*, on which Benedict's *The Lily of Killarney* would be based) on this point. With 'the taste of the English public ... yet so decidedly opposed to recitative,' Arnold would need to be 'strongly tempted indeed by the poem, and the composition, before I would venture on so hazardous and losing a speculation.... I am convinced it is by gradual and judicious advances alone, that the town will ever be brought to sanction it.'[9] By 1838, according to George Hogarth's bald statement, 'the truth is, that nobody cares anything about the dialogue of *Artaxerxes*,' and never had; it had been successful in spite of its recitative, in which, Hogarth insisted, Arne had made the mistake of setting English words in the Italian style instead of Purcell's.[10]

Towards the end of the eighteenth century, two distinct musical styles emerged within English opera. Whereas Storace, with his Mozartian connections, cultivated a cosmopolitan idiom, Arnold and Shield developed the use of national melodies. Because these folksongs occupied the place later taken by the ballad, it will be worth dwelling briefly on this feature. A continental equivalent to such use of national melodies would be Grétry's use of the Swiss *ranz des vaches* in *Guillaume Tell* (1791), though there are different cultural overtones. Concern for English music was not entirely lacking--Joseph Ritson's *A Select Collection of English Songs*, of 1783, reached a second edition in 1813--but there was more interest in Scottish and Irish material. Scottish melodies can be found in theatrical music at least from Purcell's *Amphitryon* (1690), and enthusiasm was increased by Macpherson's Ossianic literature, which appeared in the early 1760s.[11] Examples include Shield's use of a version of 'Auld lang syne' in *Rosina* of 1782 and Arnold's inclusion of 'Carolan the Irish Bard' among the composers for the pasticcio *The Castle of Andalusia* of the same year.

The 1790s saw two important collections that disseminated Scottish and Irish melodies well into the nineteenth century: George Thomson's *A Select Collection of Original Scotish [sic] Airs* (1793-99), and Edward Bunting's *A General Collection of the Ancient Irish Music* of 1796. Many of Bunting's melodies appear in Thomas Moore's *A Selection of Irish Melodies*, another influential collection, which appeared between 1807 and 1834, with part-song versions and accompaniments by Sir John Stevenson and Henry Bishop. In 1859 Novello published a selection with accompaniments by Balfe (described in the preface as 'the most popular and experienced of our native composers'), citing the tenor Sims Reeves as a particular stimulus for interest in the melodies. By this time, attention was drawn to English music by William Chappell's various publications, particularly *A Collection of National English Airs* (1838-40) and the two volumes of *Popular Music of the Olden Time* (1855-59). The accompaniments for the *National English Airs* were by Crotch, Macfarren, and Wade; those for *Popular Music of the Olden Time* were by Macfarren alone, who volunteered for the task.[12]

None of these achieved the special status of Moore's collection. The international interest in his melodies is best illustrated by the most famous example, Flotow's use of 'The Last Rose of Summer' in *Martha*, of 1847. (Moore was also known internationally as the author of *Lalla Rookh.*) In England, however, there had been a rapid decline in the use of specific Irish melodies in opera from the early years of the century. Unlike their Scottish counterparts, Irish tunes rarely have a distinguishing rhythmic idiom (the scotch snap), and since any modal elements were easily converted to tonal shapes, the melodies were absorbed into the general style of the day. Their influence was indirect, maintaining the vogue for simplicity, whereas Moore's texts, as we have seen, had a stronger influence, shaping the operatic ballad. This process of anglicization, so to speak, of melody and words divorced the musical and textual style from the unpleasant reality of Ireland; the issue of *The Times* that reviewed the first performance of *The Mountain Sylph* also included a report on 'Outrages in the County of Kilkenny'--riot, murder, mutilation--as no more than a regular occurrence.[13] Probably the last appearance of a specific Irish melody in an English opera of this phase of the nineteenth century was 'The minstrel boy to the war is gone' in Lee's 'National Ballad Opera' *Malvina* of 1826, in all likelihood the last Ossianic opera of the period,[14] and conceived as a musical mixture in the eighteenth-century style. It also includes a variant of the Scottish tune, 'Roy's Wife of Aldivalloch,' which appears in Bishop's *Guy Mannering* (1816), Davy's *Rob Roy MacGregor* (1818), and *The Mountain Sylph* (1834). But with the decline of interest in dramatizations of Scott's novels, the use of Scottish melodies also died out. Rodwell's *The Lord of the Isles* (1834) has only a few token Scottish elements, and Loder's *The Covenanters*, of 1835, and *The Gaberlunzie Man*, produced the following year, are probably the last pasticcios of Scottish tunes. (The composer-cum-arranger of the latter is unknown: the text was by William Rede.[15])

The practice of pasticcio in the broader sense of utilizing the work of several composers was supplanted at certain theatres when the necessary combination of manager, composer, and audiences' taste came into play in the 1830s, but continued where managements were less interested in music. In the early years of the century, Tom Cooke, an extraordinarily versatile instrumentalist, singer, and musical director at Drury Lane, and John Braham, the singer, were frequently involved in such pasticcios. In 1836 Cooke took advantage of his position to claim authorship of the incidental music to Planché's piece *Chevy Chace*, but the composer, the young George Macfarren, promptly withdrew the music.[16]

The important point to emerge from the above is the continuing strength of the notion of opera as little more than a play with songs.

This underlay the concept of national opera held by men as distinguished as Dickens and Bulwer. Dickens's *The Village Coquettes* (1835) allowed Hullah little room for musical effect, and as late as 1841 Bulwer could still recommend *The Beggar's Opera* to Macready as a model.[17] However, to gauge from George Hogarth's assessment, published in 1838, by this time its popularity was waning, despite being performed in a bowdlerized version: certain scenes were 'omitted, as being offensive to common decency.'[18] In accord with this moral stance, purchasers of Barnett's edition, of about 1840, were assured that it was suitable for 'the eye of the most fastidious.'

Public taste was crucial, and, expressing the traditional attitudes, was an extremely important reason for the limitations of English theatrical music in the first two decades of the nineteenth century. In Bishop's output, a loose terminology of 'musical drama' and 'opera' indicated a difference not so much of the amount of music as its distribution between characters. Works generally categorized as 'musical dramas' would include between ten and twenty musical items, not necessarily involving all the principal characters; it was in an 'opera' that the hero, as well as the heroine, would be expected to sing, but even here the number of musical sections could be limited.[19] The various pieces by Bishop, Kelly, Horn, and others normally comprised no more than solo songs of modest technical difficulty, occasional duets or trios, glees, and rudimentary finales. In the musical drama *Peveril of the Peak* (1826), for instance, Horn added to this list of ingredients only a short recitative and slightly virtuosic aria for the soprano Miss Paton. Planché recalled of the 1820s that

> Ballads, duets, choruses, and glees, provided they occupied no more than the fewest number of minutes possible, were all that the play-going public of that day would endure. A dramatic situation in music was "caviare to the general," and inevitably received with cries of "Cut it short!" from the gallery, and obstinate coughing or other significant signs of impatience from the pit.[20]

Clearly the views expressed by General Burgoyne forty years earlier continued to enjoy powerful support. The composers' stance smacks of a mixture of resignation, complacency, lack of imagination, and frustration in the face of powerful forces, among which the commercial element of sales of separate items was very strong.

Given this limited use of music, the only incentive for a foreign composer to write an English opera was financial. In no other way could it benefit his career, for, as Weber pointed out to Planché, their *Oberon* would be 'unfit for all other theatres in Europe, which is a very bad thing for me.' *Oberon* brought another difficulty to the fore: the lack of adequate performers. Planché explained that

> None of our actors could sing, and but one could act--Madame Vestris--who made a charming Fatima.... No vocalist could be found equal to the part of Sherasmin. It was, therefore, acted by Fawcett, and a bass singer, named Isaacs, was lugged in head and shoulders to eke out the charming quartuor, "Over the Dark Blue Waters."[21]

The English attitude to opera discouraged the development of high-quality singers. Hogarth maintained this had been a factor in Bishop's adaptations of Mozart ('arranged by Mr. Bishop in a form suited to the capacity of our native performers'[22]). In the first sixty-odd years of the nineteenth century, only four outstanding native singers performed in English opera with any regularity: Braham (who sang Huon in the premiere of *Oberon*), Sims Reeves, Charles Santley, and Louisa Pyne. Balfe's frequent appearances as a singer in the late 1830s reveal the paucity of talent. Though a trained singer, he was certainly not a performer of the highest ability. The use of spoken dialogue emphasized singers' lack of acting skills, all the more important when audiences were accustomed to the standards of actors such as Kean, Macready, and Helen Faucit. Planché, it will have been noted, referred to 'the play-going public.'

Despite Planché's melancholy recollections, however, it is probably from the mid 1820s that the beginnings of an increased public appreciation of music's dramatic power can probably be dated. It is difficult to estimate the contribution of Mozart's operas in this. (The earliest to reach London was *La Clemenza di Tito*, in 1806.) While highly admired by cognoscenti--Hogarth, for example[23]--for many of the general public, the effect must have been diminished by Bishop's arrangements, which, in the case of *Figaro*, left the German visitor, Pückler-Muskau, stunned, for 'neither the Count, the Countess, nor Figaro sang; these parts were given to mere actors, and their principal songs, with some little alteration in the words, were sung by the other singers.'[24] Evidently Pückler-Muskau witnessed a version that was even more diluted than Bishop's original arrangement of 1819 (described as a 'comic opera'), which included singing parts for the Countess and Figaro.

It was in a version by Bishop that *Il Barbiere di Siviglia*, the first of Rossini's operas heard in England, was given--with additions from Paisiello (1818). Rossini's visit in 1823-24 enhanced his fame, but his operas, performed at the King's Theatre, met a mixed reception. The main impact on public taste came from *Der Freischütz*, first performed in London at the Lyceum in July 1824. The acclaim was extraordinary. Fitzball, Oxenford, Planché, and Soane all produced rival versions in the autumn of the same year, for the Surrey Theatre, the English Opera House, Covent Garden, and Drury Lane. By the autumn of 1826 it had been performed over a hundred times at Drury Lane alone, with the music arranged by Bishop. On 12 April of that year, Weber directed the first performances of *Oberon* at Covent Garden; on 29 April the Drury Lane management brought out their competitor, Bishop's *Aladdin* (like Rodwell's later *The Lord of the Isles*, dedicated to the composer's mother). To an extent, Bishop bowed to the popularity of Weber's music, as witness the two-movement overture instead of the usual medley, the amount of recurrent material and of 'melodramatic' delivery of text (in the sense of speech accompanied or interleaved with orchestral passages), the inclusion of an extended movement (Mourad's scena, 'Genius of Flame!'), and the more varied scoring. But the standard English elements--the glee, the simple song, the short finales--survive. The style of the songs contrasts sharply with the more chromatic writing in the larger units, where Bishop wanted to broaden the range of musico-dramatic effects. 'Are you angry Mother?' is restricted almost entirely to tonic and dominant harmonies, with no chromaticism whatsoever. The songs are strophic, with vocal parts of the utmost simplicity--limited ranges, straightforward rhythms, short melodic units--and, except for the viola obligato played by Tom Cooke in 'In my Bower a Lady Weeps' (doubtless stimulated by Aennchen's 'Einst träumte meiner sel'gen Base' in *Der Freischütz*, but far simpler than its model), accompaniments whose piano reductions are of the same mould as those Bishop wrote for Thomas Moore's national songs, with easy left-hand octaves and right-hand broken chords. The idiom and the commercial market were well-matched. Taken as a whole, *Aladdin* is unsatisfactory not only because of its imbalance, due to the lack of dramatic arias, but because even where Bishop attempted to bring about dramatic effects the music is relatively uneventful.

Thereafter, Bishop--hitherto the dominant composer--increasingly moved aside as a composer of original operatic works, and eventually concentrated almost solely on adaptations. Of the English composers active in the late 1820s and early 1830s, only John Barnett and George Rodwell could bid to take his place, but their pieces (Barnett's *Monsieur Mallet* of 1829, for example) seem to have consisted of little more than incidental items. Two particular works may be noted: Lord Burghersh's *Catherine, or The Austrian Captive*, and Ferdinand Ries's *The Sorceress*. Burghersh (1784-1859), an amateur musician, composed seven Italian operas, all but one of which were performed privately when he was the British envoy in Florence. Among them was an Italian version of *Catherine*, a reworking of Cobb's text for Storace's pasticcio *The Siege of Belgrade* of 1791. Burgersh was the first president of the Royal Academy of Music (founded in 1822). He was highly autocratic, and during his thirty-seven years in office insisted that the operatic work should be exclusively Italian, a definition that he allowed to embrace his own compositions. In 1830, eight performances of *Catherine* were given by Academy students in the Concert Room of the King's Theatre--eccentrically, in English.[25] The Academy's staff must have been hard pressed to restrain themselves when discussing it with students. From the abysmally weak overture to each of the characterless vocal items, not redeemed by the occasional hint taken from Rossini, it was a dreadful example to set before aspiring musicians. Ries's opera is on a completely different level, as one would expect from a professional composer with a Beethovenian background. His *Die Räuberbraut* (Frankfurt, 1828) had been very popular when performed in London as *The Robber's Bride* in 1829, and to exploit this success Ries wrote *The Sorceress*, given in 1831 by Arnold's English Opera House company at the Adelphi, its temporary home after the Lyceum had burned down the previous year. The opera had a mixed reception, largely, according to *The Times*'s review,[26] on account of Fitzball's libretto (the critic loftily claimed ignorance of the author's name). The title misled the public into expecting another plot in the eerie mould of *Der Freischütz* and *Der Vampyr*, and 'great must have been their disappointment' when they discovered that the heroine, with the audience's full knowledge, merely disguises herself to defeat the conventional villain. As for Fitzball's craftsmanship, 'to enter into a critical notice of the innumerable absurdities ... would be an unpardonable waste of time and space.' Ries's music offered little compensation; though gaining some praise, generally it was considered derivative (the surviving items are in a stylistic amalgam of Mozart, Beethoven, and Weber, with a far more dramatic harmonic idiom than Bishop's), and not of a 'very high description.' After Weber, Londoners had lofty expectations of German composers.

THE ENGLISH ATTITUDE TOWARD THE RELATIONSHIP OF TEXT AND MUSIC

This is far too large a subject to be pursued in detail here; nevertheless, it is an important issue, and one that, in considering the background of the period, we should address, even though in necessarily condensed form.

The relationship of visual and verbal arts had long been explored, but, in Britain, as in various other countries, music had never been assigned a sure place as a corollary to the Ut Pictura Poesis equation. Despite a substantial body of literature on music's imitative and expressive powers, there was no consensus of opinion regarding the correspondence of music and poetry comparable to the detailed theses accepted in other fields. The system of the passions, for instance, depended on the universal identification of certain physical gestures with particular emotions. Music (especially instrumental) allowed no such widely-accepted, systematic association, even when, for want of other means of description, the terminology of the passions was applied. *The Times*'s reviewer of Ries's *The Sorceress* could say that the overture included 'a descriptive passage, in which fear, confusion, and perturbation are vividly portrayed,' presumably referring to twenty-three bars in a slower tempo and more chromatic vein than the surrounding C major allegro, but there is no evidence that this was more than an individual's response; other members of the audience may have reacted differently. The same issue arises in the case of Johann Logier's analysis of the slow movement of Haydn's quartet op.76 no.2 in his *System of the Science of Music, Harmony, and Practical Composition* of 1827. Despite his guide to the identification of the passions in music, consisting of rhythmic alterations to an unaccompanied melodic shape[27] (4.1), his reaction was a personal one, not necessarily shared by others.

The notion that music was merely 'suggestive,' and therefore allowed for the individual listener's interpretation, had been advanced in 1789 by Thomas Twining in a dissertation on musical imitation appended to his translation of *Aristotle's Treatise on Poetry.*[28] But it was typical that Twining, a scholar-clergyman and member of a wealthy mercantile family, was not a professional musician. The same applies to John Brown, James Harris, Daniel Webb, and most of the other eighteenth-century writers: apart from Avison (*An Essay on Musical Expression*, 1752), few composers expressed their views in words. Judging from their music, however, they considered the lack of a definable and exclusive correspondence between a given text and musical setting not as a weakness but an inherent advantage of music, to be exploited as eagerly as had their forefathers for century after century. When Arnold (senior) adapted Arne's overture to *Artaxerxes* as an anthem ('O be joyful unto the Lord'), for instance, he followed the practice of Handel, Bach, and countless others: re-using material, regardless of its original composer or context. Any work using the contrafactum procedure upholds the same approach. The application of new texts to existing music in George Thomson's and Thomas Moore's collections of national melodies are further examples. Moore gave a fascinating insight into this ethos in the 'Advertisement' to *A Selection of Popular National Airs* (1818-28), for which he supplied the words.

> A pretty air without words resembles one of those *half* creatures of PLATO, which are described as wandering in search of the remainder of themselves through the world. To supply this other half, by uniting with congenial words the many fugitive melodies which have hitherto had none, or only such as are unintelligible to the generality of their hearers, is the object and ambition of the present work.

The implications of this statement are too varied to be pursued fully here. We may note, however, that Moore gave no importance to the notion of a specific 'other half'; even if a melody had a particular text, this might be freely replaced by any other words, provided they were 'congenial'--a term he did not attempt to define. (Modern thought might prefer 'referential.') He would probably have considered that any such definition could only be inimical to music's powers. Nor would he have been alone in this opinion. In the words of the journalist and author Leigh Hunt, music expressed 'the exuberance of the speaker's thoughts,--the sparkles of his fire,--the numberless unutterable things which he looks and feels, but cannot define.'[29] This conviction was by no means limited to England; in essence it underlies Rossini's and

Donizetti's re-use of material from one opera to another, even though the dramatic context might be entirely different. It placed heavy responsibility on the performer's shoulders, for much of the dramatic charge of music and text was released by the singer's vocal intensity.

THE MUSICAL STYLE OF ENGLISH OPERA FROM 1834 TO 1864: THE COMPOSERS, THE APPROACH TO BE ADOPTED, AND THE SOURCES

With the above thoughts in mind, we may now turn to some preliminary matters concerning English opera from the mid 1830s to the mid 1860s.

The composers: minor figures

Over twenty composers wrote theatrical music during the thirty years under review. Many, however, such as Harroway, Lee, Lover, Romer, Smart, and Thomson, produced only one or two pieces, often with a small amount of music in each. Generally this was due to the status, managerial policy, and resources of the particular theatres with which they were associated. Until the legislative changes of 1843, the composer's task at a minor theatre was often to supply a minimal quantity of music--short passages of music to add atmosphere to a melodrama, or a few incidental songs (perhaps arrangements), in order that the resulting 'burletta' could circumvent the 'legitimate' theatres' monopoly on spoken works. If the manager favoured the traditional operatic type of a play with songs, the composer's contribution would likewise be limited. This applied to Rodwell, for instance, who was rarely able to associate himself with a musically-orientated management. Hullah's experience was similar. Even when he wrote *The Village Coquettes* (1836) for Braham's company at the St James's, the situation was unpromising, for, apart from the famous tenor-cum-manager, the singers were of limited ability, and there was an additional factor--Hullah was dominated by the personality and prestige of Dickens, who clung to the old, eighteenth-century concept of English opera.

We should note, then, that the amount of music in a work did not necessarily reflect the composer's wishes. This is clear from Barnett's 'Advertisement' to *The Mountain Sylph*. Originally, he had had no option but to plan it as a 'Musical Drama' for performance at the Victoria Theatre, which, 'owing to the want of numbers, both on the stage and in the orchestra,' would have allowed music only a limited dramatic role. He abandoned the project--not an option all his fellow composers would have been able to afford. Barnett was enabled to take up the work again, 'heightening it into an opera', by Arnold's reopening of the English Opera House at the Lyceum. Even at this theatre, however, several of the works performed were in the older mould, with little music; there was no overnight revolution, for Arnold could not afford to neglect conservative tastes.

In some cases, personal factors intervened to prevent a composer from writing more than a few operatic works. Hullah became a leading music educationalist; Thomson, who wrote the small amounts of music required for *Hermann, or The Broken Spear* and *The Shadow on the Wall*, performed at the English Opera House in 1834 and 1835, became the first Reid Professor of Music at Edinburgh University in 1838 (with Mendelssohn's support). He died three years later, at the age of thirty-six. The case of Charles Sandys Packer is unique. Packer was a twenty-five-years-old ex-Academy student when Arnold, as part of his policy of encouraging young English composers, produced *Sadak and Kalasrade* (1835). It is a more substantial work than any of Thomson's pieces, complete with passages of secco recitative in the style of Rossini's *Il Barbiere di Siviglia* or *L'Italiana in Algeri*. But Packer's criminal tendencies gradually manifested themselves, and by 1840 he had been transported to an Antipodean penal island for forgery. Later he was convicted of bigamy and found himself in an Australian prison. Emigration for a different reason influenced Henry Hugo Pearson's operatic output. In 1844 he was elected to the Reid Chair at Edinburgh, but in 1845 he moved to Germany, altered his name to Pierson, and composed for the German stage. Similarly, John Hatton's *Pasqual Bruno*, a setting of a libretto by Fitzball (after Dumas), was performed in German in Vienna in 1844, but without success. It was never given in England.

The end of the period included early efforts, all operettas, by two young composers: *Garibaldi, or, The Rival Patriots* by the precocious

Frederic Cowen, performed privately when he was a mere eight-years-old in 1860, and Frederic Clay, whose *Court and Cottage* and *Constance* were given at Covent Garden in 1862 and 1864, when he was in his twenties. Like Arthur Sullivan, they belonged to the generation who came to the forefront in the succeeding decades. (Sullivan's *Cox and Box*, with a libretto by Burnand, was first publically performed in 1867; *Thespis, or, The Gods Grown Old*, his first collaboration with Gilbert, appeared in 1871, and *Trial by Jury* in 1875.)

The major composers, and the approach adopted in investigating the musical styles of English opera in the period

As has been stated, many of the composers active during the mid 1830s to the mid 1860s contributed only a small number of theatrical works, each often consisting of a few musical items, and of which, in many cases, little now survives. It was important for us to take account of such works in the previous sections of this study, for, particularly with regard to their texts, they supply a vital element in the general picture. But to describe these as 'operas' would be misleading; generically, they are in the tradition of plays with songs. Even though in the earlier part of the century there may have been a general distinction between 'opera' and epithets such as 'musical drama,' we should beware of inferring too much from contemporary uses of the term. As we have seen, Barnett saw a clear distinction between 'Musical Drama' and 'legitimate Opera' regarding *The Mountain Sylph*, but we should also remember that Dickens consistently referred to *The Village Coquettes* as an 'opera,'[30] though the hero's musical items consist of only one song and a part in a duet and in two ensembles, and the first playbills described it as an 'Operatic Burletta.' Likewise, the theatre-goer Charles Rice referred to Rodwell's *The Sexton of Cologne* (also dating from 1836) as an 'operatic romance,'[31] whereas the publishers, D'Almaine & Co., described it as an 'opera.'

To gain an understanding of the musical nature of English opera, we shall concentrate on those works that, in addition to being set originally in English, can be defined as operas insofar as music plays an essential role in achieving dramatic effect--in characterization and the development of plot. Six composers contributed to this category: Balfe, Barnett, Benedict, Loder, Macfarren, and Wallace.

It is important to see their achievement in perspective. In terms of the basic issues of the operatic relationship between music and text, these composers sought to raise English opera to a level that not only went far beyond the boundaries accepted by Bishop in the preceding generation but had very few precedents throughout the whole of English music history. Moreover, whatever assessment of the intrinsic value of their work we may later reach, it is undeniable that, in this aspiration, they were successful.

Nevertheless, the circumstances of the day did not allow them to limit themselves to this type of opera. Barnett and Macfarren began their careers with smaller theatrical works; Loder frequently had recourse to the play with songs. Even Balfe, numerically the most successful opera composer, wrote incidental music to a play at one point in his career (Planché's *Caractacus*, 1837). Like Balfe, Benedict relied heavily on conducting as alternative employment. The publication of single vocal items, especially ballads, was another common source of income, as were arrangements of popular excerpts from foreign and English operas, for voices or piano (the operatic fantasia).

Each composer had to come to terms with the contemporary circumstances, not least managerial pragmatism. Consequently, the English tradition of spoken dialogue and the concomitant distrust of recitative are continuous elements in the period, as are the importance of the ballad and the absence of tragic endings, with their special requirements for dramatic effect. In particular, the lack of a national musico-dramatic style was crucial in determining the character of English operas. For models of broad spans of dramatic music, composers were forced to look abroad--especially to Italy and to Germany.

This was clearly recognized at the time. In 1848 Wallace wrote to James Davison, *The Times*'s critic, that 'I am at a loss when I am spoken to about an English Operatic School, never having heard of such a thing,' and spoke of drawing 'inspirations through an Italian or a German medium.'[32] Wallace's view was echoed by the music publishers Cramer, Wood & Co., when, in 1870, they advertised their arrangements of standard operas in their score of Macfarren's *Helvellyn*. Obviously the precise numbers exclude operas available only from other publishing houses (Balfe changed publishers several times), but they are significant. The list was divided into national types. Of the

'Italian School,' the company listed twenty-seven operas by six composers--Cimarosa (1), Rossini (7), Bellini (5), Donizetti (7), Mercadante (2), Verdi (5); the 'German School' was represented by seventeen operas by six composers--Mozart (5), Beethoven (1), Spohr (2), Weber (4), Meyerbeer (4; it is noteworthy that he was considered a German composer), and Flotow (1). By contrast, Cramer, Wood & Co. published only five operas by four composers of the 'French School'--Boieldieu (1), Herold (1), Auber (2), and Gounod (1). In accord with Wallace's remark, there is no heading 'English School': simply 'English Operas.' They published one each by Barnett, Benedict, Macfarren, and Wallace, and three by Balfe.

The reasons for the dominance of Italian and German opera are clear. Fundamentally, opera had always been considered Italian. As we have seen, the idea was enshrined in the 'Italian' Opera House, and in the Royal Academy of Music, long the nation's only conservatoire. (All students were required to learn Italian, regardless of their major musical interest.[33]) Italy was known as the operatic training-ground par excellence, as the cases of Braham, Pyne, Santley, and Balfe, the latter as composer and singer, show. When the Irish soprano Catherine Hayes went to study in Paris, it was not with a French teacher but an exponent of the Italian school--Manuel Garcia, brother of Maria Malibran and Pauline Viardot, who also taught Jenny Lind and Charles Santley.[34] Later, Hayes developed her professional career in Italy, rather than settling in London; Clara Novello and Anna Bishop did likewise (Novello married an Italian). For the domestic music-maker, the tunefulness of Italian opera was a strong attraction. A yardstick is provided by Chappell's publication in 1858 of *100 Operatic Melodies...*, which also includes songs and dances, for the English and German concertinas. Naturally, Chappell employed an Italian, Carlo Minasi, to compile the collections. Italian and English airs form the majority, with Verdi and Donizetti predominant. There are few French melodies; one so-called 'French Air' is actually the melody from Martin y Soler's *Una Cosa Rara* quoted by Mozart in *Don Giovanni*.

Nor did Minasi include many German melodies. By the late 1850s there was less interest in German opera than earlier in the century. Its influence was established in the 1820s, by Weber, with Ries and Marschner, of *Der Vampyr* fame, in his wake. No German composer took on Weber's operatic mantle in England during the succeeding decades. In fact, there was more operatic traffic in the other direction, for works by Balfe and Wallace were performed in Vienna, more successfully than Hatton's. Flotow's work was known in London--Bunn produced *Stradella* at Drury Lane in 1846; in a sense, however, this was a Parisian opera, having first been performed there two years previously. In the English mind, the major figure of German music was Mendelssohn, who was held in extraordinary esteem (not least by Prince Albert and Queen Victoria). But Mendelssohn had only the slightest influence in opera. He offered English composers little by way of a model, and his withdrawal from a projected collaboration left Planché disappointed and frustrated.[35] His effect was felt far more in oratorio and instrumental music, exemplified by the career of Sterndale Bennett (who seems to have done little more than toy with the idea of writing an opera), and by the founding of the Mendelssohn Scholarship, whose first recipient, in 1856, was the fourteen-years-old Arthur Sullivan. Predictably, Sullivan went to study at the Leipzig Conservatory, where his composition teacher, Julius Rietz, had been a close associate of Mendelssohn. John Barnett thought Leipzig the best place for his three children; they were fellow-students of Sullivan, overseen by Mrs Barnett. Two other composers of the later generation, Clay and Cowen, also studied in Leipzig. By the 1870s, the reputation of German music and the rigorous training was so high that Balfe's friend and biographer, St Leger, regretted Balfe had not studied in Germany, thereby acquiring 'a more serious tone.'[36]

French operas, particularly Auber's opéras comiques, gained some popularity in London, but, for all its fame, the music of French opera, as opposed to their libretti, had less effect on English composers. The clearest signs of French influence are the ballet movements that occur in several operas, such as Balfe's *The Daughter of St Mark*, for example. (Bunn employed a ballet company, as well as an opera company, at Drury Lane.) Occasional other French elements--large choral scenes, particular orchestral figurations--occur, but rarely. It was, after all, to conduct Italian and English opera, not French, that Berlioz came to London in 1847-48. The grip of the Italian and German schools was all-powerful. At the time when the English composers who would dominate the mid 1830s to the mid 1860s were forming their styles, there was no French composer whose works achieved the fame enjoyed by *Il Barbiere di Siviglia* or *Der Freischütz*. In purely financial terms, too, both for composers and singers, French opera presented far fewer international opportunities.

Cramer, Wood, & Co.'s avoidance of the term 'English School' points both to the lack of a recognized operatic tradition of international status and to the stylistic diversity of the leading English composers. It is this diversity that most characterizes the musical aspects of English opera. Individual composers chose their foreign styles and adjusted them, according to their personal circumstances and tastes, to meet the English conditions. Barnett began his career as a composer of 'legitimate opera' with a Weber-like idiom, and then adopted a wider stylistic range. Loder's development, too, was based on an essentially German approach. Balfe largely remained loyal to the Italian school in which he had been trained; Wallace, also initially an Italian in operatic spirit, gradually broadened his style. Benedict brought together German and Italian elements, adding to them a measure of Irish colour in *The Lily of Killarney*. Macfarren falls into a different category: he moved towards a revival of English traits.

We can approach the study of the music of English opera, then, in terms of the styles adopted by these composers, concentrating on their use of German, Italian, and English characteristics. For the sake of clarity, we shall focus on one composer at a time, adopting the order used above. This will give a picture of the way that each composer's musical personality and career developed, and also of the pattern of various threads that make up the history of English opera in this period.

The sources

Printed orchestral scores were very rare in England. Jackson's *The Lord of the Manor* (1780) was the last in the eighteenth century. Balfe's French operas, *Le puits d'amour* (1843), *Les quatre fils Aymon* (1844), and *L'Étoile de Séville* (1846), were printed in full score in Paris, but his London works were published in reductions for voice and piano. Indeed, no orchestral scores of any English operas were printed between 1834 and 1864. Some vocal scores give occasional hints at the orchestration (indicating a solo cello, for instance), but these are limited. The scores often consist of individual items collected together, in some cases only the most popular parts of the work.

The British Library contains a number of manuscripts of various items such as individual songs or sections of a work, but manuscript orchestral scores are, again, rare. An outstanding exception is the collection of Balfe's autograph orchestral scores and sketches, purchased from his widow. The scores include all his English operas except *Keolanthe* and *The Devil's In It*. It is difficult to be categorical about their status. The manuscript of *The Puritan's Daughter*, for instance, includes the pencilled note 'wait till Pyne ready,' a clear reference to Louisa Pyne,[37] but few other scores have marginalia relating to performance; on the other hand, Balfe's conducting score of *Fidelio* is likewise unmarked. There are several indications that the scores were used by copyists, and some pages were written by a copyist. A few manuscripts reflect the progress of Balfe's career, notably *The Bohemian Girl*, a compilation of versions spanning a period of years. The overture is different from the standard English version, and there are various sections of Italian translations of the text. The books of sketches, too, gather pages written over a number of years, in only a loose chronological order.

Sketches for *Satanella* are also to be found in the Pierpont Morgan Library, New York. Barnett's autograph full scores of *The Mountain Sylph*, *Farinelli*, and the overture to *Fair Rosamond* are in the Boston Public Library, and Loder's autograph full scores of *The Night Dancers* and *Raymond and Agnes* are in the Library of Congress, Washington. Wallace's autograph full score of *Lurline* is in the library of the Conservatoire Royal de Bruxelles. The manuscripts of Macfarren's works housed in the Fitzwilliam Museum, Cambridge, include the composer's autograph full scores of operas written before his eyesight failed completely, such as *The Devil's Opera* and *Don Quixote*.

NOTES

1. John Dennis, *An Essay on the Opera's after the Italian Manner* (London, 1706), p.14.

2. *The Spectator*, no.18 [21 March 1711]; *The Dunciad*, Book IV, line 43.

3. *The Complete Works*, edited by P.P.Howe, 21 vols (London & Toronto, 1930-34), 12:336.

4. 'On the Opera,' *The Complete Works*, 20:93-94.

5. *Some Reflexions Concerning Operas, Lyric Poetry, Music, &c.* (London, 1740), p.x. Lockman's own recitative is in heroic couplets.

6. For the following quotations see pp.x-xii, xix, xxi of Burgoyne's preface.

7. Here meaning the type of instrumental prelude or ritornello that regularly occurred in songs.

8. See Bruce Carr, 'The first all-sung English 19th-century opera,' *Musical Times*, 115 (1974):125-26.

9. Daniel Griffin, *Life of Gerald Griffin Esq. by his Brother, Daniel Griffin, M.D.* (London, 1843), p.212.

10. *Memoirs of the Musical Drama*, 2 vols (London, 1838), II:89-90, repeated in the second edition, *Memoirs of the Opera in Italy, France, Germany, and England*, 2 vols (London, 1851), 2:48-49.

11. The effect of this enthusiasm was by no means limited to operatic works, as J.C.Bach's use of 'The Yellow Hair'd Laddie' in a keyboard concerto (op.13 no.4, of 1777) indicates.

12. See Chappell, *Popular Music*, 1:xi.

13. *The Times*, 26 August 1834, pp.5, 6.

14. Eric Walter White followed Allardyce Nicoll in wrongly attributing *Malvina* to Cooke (and Blewitt's *Auld Robin Gray*) to Lee. See White, *A Register of First Performances of English Operas and Semi-Operas* pp.26, 28; Allardyce Nicoll, *A History of English Drama 1660-1900*, 6 vols (Cambridge, 1952-59), 4:349. The printed score, libretto, and playbills are quite clear about the names.

15. The playbills advertizing the performances at the English Opera House (Adelphi) merely state that the music was 'selected entirely from favorite [*sic*] Old Scottish Melodies.'

16. Henry C. Banister, *George Alexander Macfarren: His Life, Works, and Influence* (London, 1891), pp.39-40.

17. Charles H. Shattuck, ed., *Bulwer and Macready* (Urbana, 1958), p.203.

18. *Memoirs of the Musical Drama*, 2:55.

19. See Bruce Carr's account in *The Romantic Age*, pp.292-93.

20. Planché, 1:80.

21. Planché, 2:81-82.

22. *Memoirs of the Musical Drama*, 2:385.

23. *Memoirs of the Musical Drama*, 2:225-67.

24. Pückler-Muskau, 3:147.

25. Rev. W.W. Cazalet, *The History of the Royal Academy of Music* (London, 1854), pp.248-50.

26. *The Times*, 5 August 1831, p.5.

27. Johann Logier, *System of the Science of Music, Harmony, and Practical Composition* (London, 1827), pp.319-20, 266-67.

28. Thomas Twining, *Aristotle's Treatise on Poetry, translated: with two Dissertations...* (London, 1789), p.49, fn. s.

29. Leigh Hunt, *Musical Evenings*, edited by David R. Cheney (Columbia, 1964), pp.34-35.

30. See, for example, Dickens's letter to J.P.Hullah, [?19 July 1836], *The Letters of Charles Dickens*, The Pilgrim Edition (London, 1965--), 1 (edited by M. House & G. Storey): 154-55.

31. Rice, *The London Theatre in the Eighteen-Thirties*, p.38.

32. From a letter in private possession, quoted by White, p.290.

33. Cazalet, p.336.

34. Garcia moved to London, and from 1848 to 1895 taught at the Royal Academy of Music.

35. Planché, 1:279-316.

36. Heyward St Leger, *Reminiscences of Balfe* (London, [1871]), p.x.

37. BL, Add.29359, f.89.

V

Barnett

Career and Output

John Barnett (1802-1890) was articled to Arnold, the manager of the English Opera House, as a boy alto. His other teachers included William Horsley, an organist and composer of vocal music, the singer and operatic composer C.E. Horn (the son of C.F. Horn, a prominent teacher), and the German musician Ferdinand Ries, who had been associated with Beethoven. From 1826 onwards Barnett composed incidental music for a large number of stage plays, and in 1832 became musical director at the Olympic theatre, then managed by Madame Vestris, the original Fatima in Weber's *Oberon*.

We may gauge the frustration that Barnett must have felt during these years from the advice offered by the playwright John Buckstone when dedicating *The Pet of the Petticoats* to the composer. First performed in 1832, this 'Opera in Three Acts' was revived in June 1834, not long before the premiere of Barnett's *The Mountain Sylph* at Arnold's English Opera House on 25 August.

> That you are capable of productions of a higher order than the opportunities this Opera afforded, all who know you are ready to believe; and in the present neglect of English musical talent, you are right to hide your candle under a bushel.
>
> In the mean time take a hint from one of your well-wishers. As you spring from a German stock,[1] go to Germany--change your name--be no longer John Barnett, but VON *Somebody*,--write an Opera, produce it at one of the German theatres, it will surely be brought to England,--

> become popular,--its music will be whistled in the street, warbled in the drawing-room, and the new German composer will be immortalized! Do this and shame us!

Buckstone's remarks point to the lack of opportunities available to English composers and, in the wake of Weber's success, the acclaim given to German musicians. Equally obvious is the importance of Arnold's 'spirit, zeal, and enthusiasm ... in the production of Native Operas, and in cherishing native talent,' to quote from Barnett's dedication of *The Mountain Sylph* to the manager.

This opera was recognized as an important and influential work, but Barnett's subsequent career did not fulfil its promise. Two further operas were performed at Drury Lane--*Fair Rosamond* (28 February 1837) and *Farinelli* (8 February 1839). In 1841, having failed in his own attempt at theatre management, Barnett settled in Cheltenham. By this time he had offended Arnold, the critic Leigh Hunt, Balfe, and many others.[2] *Kathleen* and *Queen Mab*, written at the time of his move from London, remained unperformed, and *Marie*, of 1845, was never completed.

In this discussion we shall focus on *The Mountain Sylph* and *Fair Rosamond*.

The Mountain Sylph

The Mountain Sylph--'my first attempt at legitimate Opera,' as Barnett described it in his *Advertisement*--holds a special place in the history of English opera. It broke from the norm cultivated by Bishop and others in that music plays an integral role in characterization and in important action, notably in concerted pieces.

The work does not, however, entirely break with tradition. Speech is used extensively, and one of the main characters, Christie, sings very little. There is only a small amount of recitative. The Scottish setting--on the Isle of Skye--and the inclusion of Scottish melodic material were also familiar aspects of English theatre music. 'Roy's Wife of Aldivalloch,' long a favourite melody, appears in Donald's (the hero's) second aria ('Art thou a form of mortal birth'), though, unusually, Barnett allocated the Scots tune to the accompaniment (5.1). Donald's first solo, 'Oh, no, 'twas no deceptive spell,' may not be from an existing song, but, with its scotch snaps, it is imbued with the flavour of Scottish melodies (5.2).

Another standard ingredient is represented by the three simple strophic songs in Act II. 'Can'st thou love, yet coldly fly me?' and 'Thou art not he, whose looks of love,' for Donald and Jessie, the second female role, were inserted after the libretto was sent to the Lord Chamberlain's office. Doubtless they were included at the request of the singers, Wilson and Miss Somerville. Both are conventional for the time: the duple-time melodies are divided into regular, short-breathed phrases, sometimes with longer notes matching the ends of verbal lines. However, the 3/8 song 'Farewell to the mountain,' written for Henry Phillips (whom Barnett especially thanked in his 'Advertisement'), is more in the style of a Wiegenlied--a German lullaby (5.3).

The Lord Chamberlain's copy of the libretto contains the texts of a song and an extended aria for the heroine, the Sylph Aeolia, which do not appear in the vocal score or printed libretto.[3] Presumably they were excluded from the opera in favour of the songs interpolated for the three singers noted above--a decision that cannot have been welcomed by Miss Romer, a soprano who enjoyed a substantial career (unlike Wilson and Miss Somerville). Barnett made no alteration in his later edition of about 1880, but on occasion Aeolia's part was made more weighty by interpolating 'Non più mesta' from Rossini's *La Cenerentola*, in its original key of E major, immediately before the final C major chorus. This is evident from markings in the vocal score in the library of the Royal Academy of Music, once owned by J.W.Thirlwall, a publisher active in the 1840s-1860s. Presumably it was sung in English.

The opera's conventional features are offset by its exceptional aspects. The source of the plot--the earliest of the outstanding romantic ballets, *La Sylphide* (first performed in Paris two years earlier)--is symptomatic of a new era. Donald is engaged to marry Jessie, but is secretly in love with Aeolia. This is revealed by the malevolent Hela. Donald pours scorn on the villain, but during the marriage ceremony he disappears with Aeolia. Plotting revenge, Hela summons his demonic witches in an Invocation scene (the Act I finale). His soliloquy, whose text was discussed in chapter 2, leads to an attempt to kill Aeolia. She is saved by the Sylphid queen, who finally sanctions the love between Donald and Aeolia, transforming her into a human. The opera ends with general rejoicing that 'faithful love still is guarded from above.'

The influence of Weber's *Der Freischütz* is clear from various characteristics of the opera, not least the use of recurrent motifs associated with elements of the drama. The overture is not fully synoptic, for it excludes the heroine's motif and concludes with those associated with evil, but the model is obvious. Satanic overtones are announced at the outset by means of the diminished-seventh chord. Tonal stability is avoided to a noteworthy extent, due to Barnett's association of a clear sense of tonality with virtue and its antithesis with evil. This notion stems from Weber's example, but the Englishman used it to particular effect: the tonally-twisted melody to which the witches sing 'Charming sister, here we've rid from sulphur caverns merrily, merrily' in the Act I finale is extraordinary (5.4).

As in *Der Freischütz* the rural community is regarded as morally benign. Its musical character is established by a bridal chorus (5.5.a) which includes an orchestral figure in scotch-snap rhythm. It is an inversion of the second bar of 'Roy's Wife of Aldivalloch' (5.5.b). One of the motifs linked to Hela is a derivative of this. It provides an ostinato-like accompaniment during his long solo 'To me what's mortal happiness'[4] as he recalls the time when he, too, was a member of the benevolent community: 'In days gone by/ Alas, was I,/ a flow'ret gay as rare' (5.6.a). When infected with diminished-seventh harmonization this motif signifies his malevolent nature--innocence perverted to evil (5.6.b). The diabolic element is made explicit when the motif reappears in the Act I finale (5.6.c). The latter includes two other figures drawn from Hela's solo, a further diminished-seventh figure in arpeggio shape (5.7), and a scale pattern for 'Man's misfortune glads my soul' which, while not using the diminished seventh, is tonally unstable. This passage was remarked upon by *The Times* as 'descriptive of vindictive passion'[5] (5.8).

'To me what's mortal happiness' is the most expansive of the solo items. It is a 'scena,' a term that, in English opera, denotes a complex corresponding to Agathe's 'Wie nahte mir der Schlummer' in *Der Freischütz*. The conventional English scena consists of four sections. The first comprises recitative and measured dramatic declamation over varied orchestral patterns. Thereafter come a slow lyrical movement, an interlude, and a fast movement. Barnett used this sequence to reflect the textual disclosures regarding Hela's personality by means of sections characterized by distinctive musical figures, a fastidious attitude that owes much to Weber's example. The use of recurrent material has been noted above. The first section is based in C minor but is tonally unsettled. As Hela dilates on his past virtue ('In days gone by') the key stabilizes in E flat. The interlude is set in C minor, preparatory for the concluding allegro. The latter was foreshadowed in the overture. It opens in a blaze of C major--'Lightnings flash and thunders roar'--a perverse celebration of triumph that is undermined by the ensuing tonal instability. At each stage the scena reveals a deep concern to correlate music and text and an unusual quality of invention.

The major solo in Donald's part is 'Art thou a form of mortal birth.' The preceding recitative is old-fashioned in its style (5.9), but the aria is an original conception. Its form is moulded by the dramatic situation. Donald's fears about contact with the supernatural Aeolia are represented by a disruption of tonal stability and lyrical flow by a passage of arioso over tremolo accompaniment. The dissonant harmonies contrast starkly with the diatonic idiom and pedal bass of the Scottish melody. The opening phrase returns, but for less than four complete bars: Donald excitedly calls upon the Sylph to appear, and to fortissimo horn chords, soft flourishes from a solo flute, and tremolo harmonies, Aeolia makes her appearance. Recitative is succeeded by measured declamation with orchestral patterns that change according to the text. Donald is again fearful, and the accompaniment portentously outlines diminished sevenths: but Aeolia interrupts him and the 6/8 recurrent material associated with the Sylph, already heard in the previous recitative, appears in the accompaniment (5.10). Moving with an unusual degree of chromaticism, Donald's part finally settles on the harmony that introduces Aeolia's aria, 'Deep in a forest dell,' whose first phrase consists of her motif.

The Times noted that the opera 'is chiefly rich in concerted pieces, many of which are distinguished by a true dramatic character.' This implies more such pieces than is the case, but important junctures in the plot are indeed set to music. These include the fortune-telling episode and the marriage-contract scene in Act I. The former characterizes the individuals distinctly, and Donald's outburst against Hela is vividly portrayed with chromatic writing and frenetic orchestral figures (5.11). By contrast, when Jessie asserts Donald's reliability in the subsequent trio the tonality is established by cadences that are almost Mozartian (5.12).

Both acts end with extended finales. The first discloses much of Barnett's technique. Its opening sombre harmonies contrasted with loud dissonant chords and the use of tremolo strings announce his debt to the Wolf's Glen scene in *Der Freischütz*. Similarity with the German example is increased by the use of 'melodramatic' delivery--speech with orchestral accompaniment--and the recall of motifs, here taken from Hela's scena. These are interspersed with the witches's material noted earlier, including dancing rhythms to highlight their malevolent glee ('Shall we curse or shall we kill, or merely torture merrily, merrily').

Barnett was unable to match his exemplar in sustaining tension--a weakness that arises at various points in the work. Nevertheless *The Mountain Sylph* represents a considerable achievement. Despite obvious reliance on various models, there is much originality. But the work's importance goes beyond this. Flying in the face of long-held assumptions, it demonstrated the viability of English opera as a dramatic genre.

Fair Rosamond

Fair Rosamond illustrates Barnett's work on a larger scale. *The Times* reported that the premiere lasted four and a half hours--reflecting Bunn's love of the grandiose. It will be recalled that this is one of the few operas of the period to be set in England. King Henry II has developed an adulterous affair with Rosamond, who is unaware of his true identity and that he is married to Queen Eleanor. Only at their coronation does Rosamond realize the truth. The Queen discovers Rosamond's hiding-place and is about to kill her when Henry arrives. Rosamond pleads on Eleanor's behalf, and all are reconciled. *The Times* was highly critical: Rosamond emerges 'not as a woman to sin no more, but as a spotless angel.... The poor composer never could have contemplated such a conclusion.'[6] The last remark was an obvious exaggeration, but it made the point unequivocally.

Musically, the opera is of quite a different ilk from *The Mountain Sylph*. There is a mixture of styles that reflects the insecure basis of English opera at the time. Weber's influence is reduced to the use of recurrent melodic phrases first heard during the overture and to the scenas for Henry and Rosamond. Textually descended from Agathe's main solo in *Der Freischütz*, Rosamond's 'Ah, me, he comes not' is more clearly divided into movements than Hela's, but in Henry's 'The morning breaks' only the second movement ('Sweet sleep') approaches formal regularity; the others are freer. The initial orchestral introduction is marked by a contrapuntal texture and chromatic part-writing that set Barnett apart from his English colleagues. The recitative culminates in a highly unusual flourish, echoed by the obligato viola that is prominent in 'Sweet sleep' (5.13). This scoring would have been suggested by 'Einst träumte meiner sel'gen Base' in *Der Freischütz*, already imitated by Bishop in *Aladdin* ('In my Bower a Lady Weeps'); the result is close to Schubert's song 'Viola' (5.14). Barnett's melody, however, is characteristic of the composer in its triadic leaps, repetitive rhythmic patterns, and a structure that includes a dominant close in the eighth bar--reflecting a conservative taste in melodic construction. The scena continues with two further lyrical movements, slow and fast, whose restrained textures and harmonic styles are sharply differentiated from the chromatic harmony and energetic orchestral figures of the dramatic passages (5.15). The entire solo is an imaginative piece, fluid in form and resourceful in its effects, testifying to Barnett's preoccupation with musico-dramatic expression. At the other extreme of musical style, the folk-like melody sung by Henry when disguised as a minstrel ('Edgar') in Act I is another outcome of this concern (5.16).

The influence of the grand French style is apparent in the ballet movements in Act IV and in the large choral tableaux, especially the finales. The dramatic situation in the Act III finale is so close to that of the Act I finale of *La muette de Portici* (the unwitting mistress discovers her lover's identity at a public ceremony) that inevitably there are similarities. Barnett's anthem (5.17) corresponds to the choral prayer that precedes Auber's finale. But the trio 'Wild my brain to phrenzy driven,' marked 'canone' (5.18), is closer to the style of a Rossinian ensemble 'di stupore' than Auber's 'Quel est donc ce mystère?'. Of yet another idiom is the canonic quartet in the Act II finale (5.19). Here the model of 'Mir ist so wunderbar' in Beethoven's *Fidelio* springs to mind. On the other hand, the concluding hemidemisemiquaver flourishes recall Rossini's fioriture.

The five-part madrigal in Act IV ('Merrily wake music's measure') is an element derived from the English style of Bishop, for it is a glee with an antiquarian gloss. Barnett's use of national musical material--the 'Morrice Dance' in the finale of Act I, and 'Edgar's' folk-like melody--further relates this work to earlier phases of English opera. As in *The Mountain Sylph*, however, this characteristic music is used more thoughtfully than by previous composers.

Regarding the solo songs, Barnett made less concession to the commercial marketplace than in *The Mountain Sylph*. We may point to Rosamond's 'Mine must be the silent tear' for its affecting quality of restraint (5.20). Henry's 'Ballad' 'That smile so sweet' is unique in this period of English opera. In order to reflect the poetic metre, the time signature alternates from 3/4 to 4/4 during the first eight bars (5.21).

The strength of the opera derives from the composer's dramatic imagination. The Queen's attempt on Rosamond's life in the Act IV finale is an example. Barnett contrasted a lyrical strain for the heroine with a repeating bass pattern for her exchanges with the Queen, a particularly individual technique (5.22). Thereafter the finale soon falls flat, but, as we know, this was not Barnett's fault. He included one especially nice touch: when Rosamond begs Henry to forgive the Queen, she does so to the strains of the minstrel melody he had used to woo her in Act I.

SUMMARY

Barnett impresses as a highly individual composer blessed with a strongly-developed dramatic sensibility. At its best, his work is marked by a vivid responsiveness to the text and an imaginative use of musical resources. Indeed, he is most convincing when most inventive; the scenas for Hela and Henry are unmistakable witnesses. He took great pains to illustrate the human passions, and in this regard he was exceeded by no other English composer of the period. One consequence is that his use of tonality for dramatic purposes is the most unusual to be encountered in the years under review.

The Mountain Sylph set new standards as regards the composition of dramatic music. But to limit an assessment of Barnett's contribution to the development of English opera to this work does him less than justice. It must be admitted that *Fair Rosamond*'s stylistic diversity robs it of a sense of unity, emphasized by its unwieldy size. Nevertheless it reveals a versatile composer, intent on sharpening his portrayal of dramatic effects. In its English setting and use of indigenous musical material the work provides a historical link between the operatic style of the late eighteenth and earlier nineteenth centuries and, as we shall see in due course, Macfarren. It shows the direction that one talented, if eccentric, English composer might have pursued had circumstances been different.

NOTES

1. The composer's father was a Prussian merchant who emigrated to England, altering his surname from Beer to Barnett. He is said to have been related to Meyerbeer.

2. See *The Musical World*, 18 (1843): [371]; 396, 403, 429-30; E.D. Mackerness, 'Leigh Hunt's Musical Journalism,' *Monthly Musical Record*, 86 (1956): 219-21; Kenney, pp.153-55.

3. See BL, Add.42927, ff.470v-71, 476-76v, and the 'second edition, corrected' of the *Songs, Duets, Choruses, &c. in The Mountain Sylph*.

4. 'To' was misprinted as 'For' in the vocal score published by W.Hawes in 1834.

5. *The Times*, 26 August 1834, p.5.

6. *The Times*, 1 March 1837, p.6.

VI

Loder

Career and Output

Like John Barnett, Edward Loder (1813-1865) was a pupil of Ferdinand Ries, with whom he studied in Frankfurt am Main from 1826 to 1828. His operatic career may be dated from the premiere of *Nourjahad* on 21 July 1834, being the first new opera produced at Arnold's rebuilt English Opera House. It is a small work, noteworthy for its indications that the young composer was already especially interested in emphasizing dramatic effects. In a duet for Schemzeddin and Nourjahad, for instance, a single interjection by the latter merits an A major chord in the midst of a passage in flat keys. This type of imaginative and detailed response to a dramatically-charged situation lies at the heart of Loder's best work. Years were to pass before he could develop this early potential. With the collapse of Arnold's enterprise Loder was forced to enter a contract with the publishers Dalmaine & Co. to produce a new composition every week. His theatrical work was reduced to adapting and composing small amounts of music, particularly songs, some of which were collected to form *Francis I* of 1838, his only work to be performed at Drury Lane. In 1846 he was appointed musical director at the Princess's Theatre, where *The Night Dancers* was performed on 28 October of that year. He remained in London until 1851 when, with John Maddox's management at an end, he moved to the Theatre Royal, Manchester. The following year Berlioz conducted *The Island of Calypso*, in effect a cantata, in London.[1] Four years elapsed before Loder's grandest opera, *Raymond and Agnes*, was produced in Manchester (14 August 1855). By the end of the year he was suffering from a cerebral disease and in time returned to London, where *Raymond and Agnes* was given at the St James's Theatre on 11 June 1859. It was shortly followed by *Never Judge by Appearances*, an operetta composed in Manchester (the libretto was by Drayton, who created the Baron's role in *Raymond and*

Agnes), at the Adelphi on 7 July. Loder was increasingly paralysed. The last four years of his life were spent in a coma.

Loder's theatrical output includes four works that remained unperformed. His vocal music embraces a number of glees and songs, most of which were undisguisedly adapted to the market place. His settings of 'The Brooklet' (whose text is based on Müller's 'Wohin,' set by Schubert in *Die Schöne Müllerin*) and of sacred texts are more truthful indications of his ability.

The present discussion concentrates on *The Night Dancers* and *Raymond and Agnes*, Loder's major operas.

The Night Dancers

As we have seen in chapter 3, Soane's source for the plot of *The Night Dancers* was a German legend. On the eve of her wedding, the unsophisticated Giselle dreams that her beloved Albert is actually a prince, thus dashing all her hopes of happiness. Albert's father and his fiancée, Bertha, seek him out, and his true identity is revealed by the inquisitive Fridolin. The heroine is unhinged by these events and soon expires. Now transformed to a supernatural being, she is joined by the Wilis--female spirits who dance unfaithful men to their deaths. The distraught Albert summons Giselle's spirit. Like two other prominent nineteenth-century heroes, Heathcliff in Emily Bronte's *Wuthering Heights* (1847) and Wagner's Tristan, he wishes for death in order to join his beloved. Giselle saves him from this fate, asking only that he will not forget her. At this sad juncture the heroine awakens to find all was a nightmare, and her marriage to the trustworthy Albert may proceed.

The Night Dancers was influenced by practical issues. Maddox engaged a well-known soprano, Emma Albertazzi, to sing Giselle's role. Her status influenced the structure of the opera, for the heroine's part includes seven solos as opposed to three for Albert. As a whole the work is constructed on a small scale, reflecting the modest resources Maddox could otherwise command. There is a strong element of music tailored for public consumption--predictable in the light of Loder's career--resulting in simple strophic songs with an unvarying rhythmic pattern such as 'I cannot flatter if I would' (6.1) and glee-style choruses. Indeed, the printers' plates of the 'Lake Serenade,' scored for alto, two tenors, and bass, were marked 'The Boat Glee.' The integral role played by dance in the plot satisfied both commercial and dramatic interests. The operatic use of popular dances was frequent. In 1843 Balfe had incorporated a waltz and galop in *The Bohemian Girl*, for instance, and to these Loder added a tarantella, familiar in opera at least since Auber's *La muette de Portici* (1828). Waltzes supply the background for Giselle's aria 'Wild is the spirit' and the final movement of her scena 'I dreamt we stood before the Altar,' which recurs as an independent instrumental piece. In the vocal score the Act I finale includes a 'Mazurka or Silesian Peasants' Dance' attributed to 'Miss and Mr Marshall,' as well as a 'Silesian Waltz' specified as being Loder's work.

Loder's German training left an indelible mark. The essential stylistic character of *The Night Dancers* stems from Ries's greater contemporary, Weber. The opera opens with a humorous narrative ballad which is the equivalent of 'Einst träumte meiner sel'gen Base' in *Der Freischütz*, and the use of 'melodramatic' delivery also derives, ultimately, from the German style. More specific parallels occur in Giselle's 'Grand Scena' 'I dreamt we stood before the Altar.' Though this type of solo was by now a familiar feature in English opera (precedents for heroines included those in Barnett's *Fair Rosamond* and Balfe's *The Daughter of St Mark*, of 1837 and 1844), it may also be traced to German opera. The opening triplet figure is very similar to that used by Rossini in *Guillaume Tell* (6.2), but at other points Agathe's scena in *Der Freischütz* was clearly the model. Soane obviously had the German text in mind: compare 'The full Moon shone round and bright' with 'Ob Mond auf seinem Pfad wohl lacht? Welch'schöne Nacht!'. Loder's harmonies are similar to Weber's (6.3). The interlude between the slow and fast movements adds correspondences. 'E'en while I gaze' has a clear resemblance to 'Dort klingt's wie Schritte' (6.4), and the care taken over orchestral effects bears witness to Weber's example.

Albert's recitative ''Tis madness' and the following aria 'Wake from thy grave, Giselle,' however, are written in a synthesis of styles. It is typical of Loder that the former is more a lyrical arioso than a recitative (6.5). It has the atmosphere of Max's words of despair in *Der Freischütz*, 'O dringt kein Strahl durch diese Nächte,' and, as did Weber at the comparable point, Loder incorporated recurrent material (first sung by Giselle shortly before her death in the Act I finale). When Albert calls upon Giselle's spirit, the tonality changes by means of an enharmonic device Loder might first have encountered in Ries's *Die Räuberbraut* of 1828 (6.6). The opening of the aria is also characteristic of Loder's work in reaching a high pitch early in the melody and in falling into units matching the lines of text at the cost of a long melodic line. But in the more emotional central section the language becomes reminiscent of Bellini and Donizetti, with frequent appoggiaturas, syncopations in the vocal line, and diminished sevenths approaching cadences (6.7).

By contrast, Albert's strophic song 'Wake my love all life is stirring' (6.8) owes little more than its cadential fioriture to the above composers. It includes some of Loder's most affective and individual melodic writing.[2] Indeed, his most overtly Italian arias are the least successful. This applies to Giselle's 'What pleasure can there be in slaying,' marked by Rossinian semiquaver passages, and in the Duke's 'For one sweet hour, alas, too brief,' with its phrasing, dotted rhythms, minim-length appoggiaturas, and roulades, familiar from the pages of *Lucia di Lammermoor* and *La Sonnambula*--compare, for instance, Amina's 'Ah! non giunge uman pensiero' from Bellini's opera (6.9). Albertazzi may well have demanded the former solo; as we have already seen, certainly the latter was written at the request of the singer, Bodda. It is significant that the recitative preceding Giselle's aria ('I breathe again') concludes more in Mozart's than Bellini's or Donizetti's vein--it is similar to Donna Anna's 'Troppo mi spiace' in *Don Giovanni* (6.10). The comic Fridolin is rarely characterized by means of Italianate buffo idioms.

The concerted pieces are weak. The longer of the two, 'A noble train,' opens with orchestral figuration that too obviously suggests that the instrumental music was composed first and the voices then superimposed. Likewise, the septet 'Wilt please you stay with us a while' gives the impression that Loder wrote a simple melody, divided it between the two sopranos, and harmonized it in appropriately static fashion. The substantial solo for Bertha is so limited in compass and rhythmic variety as to lower the effect of the whole.

By contrast, the challenge of setting Giselle's death in the Act I finale stimulated Loder to some of his finest work. He reduced the amount of musical activity to increase the pathos of the sombre event. At an almost hymnal pace the compassionate Bertha offers Giselle her engagement ring; Loder makes us focus on Giselle's despair by repeating her earlier theme 'The Prince! then I indeed am lost' in the orchestra. The heroine's final moments are charted in the harmonies, which are so fluid that the key signature is dispensed with. Giselle quietly recalls a motif from her love duet with Albert, 'He loves me, loves me not,' above major chords, but Bertha and Albert respond with dissonant progressions. Giselle's motif is heard on solo woodwind instruments over tremolo string chords. 'Rouse, love; rouse, love; shake off this counterfeit of death,' vainly coaxes Albert, now with major sonorities. To the fateful tolling of a bell and the sound of a diminished-seventh chord--imaginatively scored for bassoons, horns, trombones, and timpani--Giselle breathes her last. It is an impressive and moving scene (6.11).

The quality of such writing emphasizes the inconsistency of *The Night Dancers*, not only in terms of the varying styles Loder used but also because of his fluctuating standards of composition. The items written to order, so to speak--those aimed at the amateur musician browsing among the latest songs, or supplied to a singer's specification--are the weakest. It was when confronted by a vivid dramatic situation, as in Giselle's scena and her death-scene, that his strengths came to the fore.

Raymond and Agnes

Raymond and Agnes is Loder's only opera conceived on a scale comparable to those of his contemporaries such as Balfe and Wallace. It portrays a wider range of characters and is musically more continuous than any of his other theatrical pieces. The work has a quality of invention and dramatic power that raises it to an unusual position in English nineteenth-century opera.

It was selected to open the season given by Augustus Braham's company at the St James's in June 1859. The choice was probably influenced by the tenor, George Perren, who had played Raymond in the Manchester cast, and the conductor, the composer's cousin, George. (Augustus was a son of the famous John Braham.) Some revisions, involving a reduction from four to three acts, were made for the new production. The following study is based on the revised version.

To place the work in context we may recall that the London performances took place a quarter of a century after Loder's first opera, *Nourjahad*, had appeared, and thirteen years after *The Night*

Dancers. The most recent English operas given in London were Balfe's *The Rose of Castille* and *Satanella*, performed by Louisa Pyne's and William Harrison's company in 1857 and 1858. Due to the summer season, Braham's competitors were the rival Royal Italian Opera companies, one performing *Don Giovanni* and *Otello* at Covent Garden (with a cast including Grisi and Mario), and the other *Il Giuramento*, by Mercadante, and *La Traviata* at Drury Lane. Competition also came from the Handel Festival at the Crystal Palace. In these circumstances high standards of performance were all the more important, but Chorley reported that the first night 'was better attended than performed.'[3] Perren and Hamilton Braham, Augustus's brother, who played the villainous Baron of Lindenberg, were far weaker than the women singers. The part of Agnes was performed by Madame Rudersdorff, a soprano with some experience in Italian opera,[4] and that of Madelina by the mezzo-soprano Susan Pyne, sister of the more illustrious Louisa. The remaining female role (Ravella, mute for most of the work) is a speaking part. *The Times* allowed less space for its review than for a report of the recent Oxford ceremony in which young John Stainer received his Bachelor of Music degree.[5] Performances ceased after a week.

The opera suffers from a poor libretto. Fitzball's plot is an uneasy rendition of an episode from Lewis's novel *The Monk*. The young Spanish nobleman, Raymond, is in love with Agnes, the ward of the Baron of Lindenberg, who himself intends to marry her. Raymond's discovery that it was Lindenberg who, years previously, killed his father and kidnapped his mother leads to his imprisonment in the Baron's castle. The villain's remorse is heightened by the spirit of the heroine's holy namesake, St Agnes. Raymond and Agnes take advantage of this to escape, but unwittingly seek succour from a gang of robbers led by Antoni, once the Baron's lieutenant. The Baron and his men recapture the couple, and Lindenberg plans Raymond's death with his old henchman. Agnes vainly searches the castle for her beloved; as if to reassure the heroine and audience, there is revealed a mystical vision of her marriage to Raymond, attended by St Agnes and a choir of nuns. Eventually it is Lindenberg who, mistakenly, falls victim to Antoni's bullet. The latter's dumb slave, Ravella, regains her power of speech and identifies herself as Raymond's long-lost mother. The opera concludes with a celebration of Providence's eternal concern for the innocent.

The handling of events is often unconvincing, and the treatment of the comic personality (Theodore) is incongruous. Fitzball inserted stock conceits for the hero and heroine in unsubtle fashion. Thus Raymond's ballad 'Farewell the forest and the plain' includes the topoi of isolated hero, woman as angel, parental home, fidelity that allows death to be faced contentedly, and a tear as a proof of love.

Various aspects of *Raymond and Agnes* point to the strength of Loder's imagination. Even so, the opera includes certain items intended for the domestic market, the bane of Loder's career. The most obvious instances are the uninspired song written for Susan Pyne, 'Oh well do I remember,' and 'Where the pearly dewdrop falleth,' a glee complete with 'tra la' refrain. However, the extremely elementary type of song exemplified by 'I cannot flatter if I would' in *The Night Dancers* was excluded in favour of more modern ballads.

Loder's basic affinity as a musical dramatist remained with Weber. The overture, for instance, adumbrates the fundamental issues of the plot. A short dramatic passage gives way to the first of two prayers sung by the heroine to St Agnes ('Oh Agnes! martyr fair': compare 6.12.a and 6.24). Thereafter the focus is on the conflict of two themes. One is similar to a motif heard in the duet for the villain and hero at the point when past events are explained (see 6.12.a and 6.12.b). The other, more important in the opera as a whole, directs attention to the saint and, by association, Agnes, as is made clear in Act II (6.13). This theme recurs during the finale of Act II ('Oh! holy spirit, who thro' these gloomy tow'rs') and in the Act III finale, most pertinently at the death of the villain. The 'melodramatic' use of speech and the inclusion of a narrative ballad for the second woman again stem from *Der Freischütz*. In other instances Weber's long-term influence had been absorbed to the point where there is a close parallel of aims and methods, not direct imitation. The care lavished on orchestral figuration to heighten the effect of the text, for which the older composer was justly famous, is now a basic principle of Loder's work.

As regards the scenas, we may first consider the Baron's 'Madrid, Oh! Madrid.' Certain features are shared with Giselle's equivalent in *The Night Dancers*: the basic form, the key scheme (fundamentally E minor--G major--E major in each case), and the nature of the andante's generally smooth melodic line and slow-moving accompaniment. As in Giselle's case, too, there is some reliance on an Italian orchestral figuration to add expressive weight to the introductory movement, here an appoggiatura figure emphasizing the villain's guilty distress (6.14.a).

A comparable pattern has already appeared in the Act I finale in a similar context, where the Baron cries 'It is remorse' (6.14.b). Characteristically, it is not primarily by means of mellifluous melody that Loder achieved his end but by using dramatic figures involving voice and orchestra, as is witnessed by Lindenberg's outburst in the andante of the scena (6.15). The allegro, 'No cypress o'er my tomb shall wave' opens with one of Loder's best melodic inventions, an arching four-bar phrase in E minor (6.16). The succeeding section does not maintain this quality--the words 'Gone for ever hope of bliss' are soon set to a major key (settling in E), revealing a compromise between the mood of the text and the standard tonal structure. Nevertheless, taken with the powerful expression of his remorse in the first finale, these illustrations of the Baron's personality result in one of the most rounded portrayals of a villain in any English opera of the thirty years now under review.

Agnes's 'In vain I wander' is particularly interesting because Fitzball supplied the materials not for a scena but a four-line recitative followed by a ballad whose two verses are full of conventional images for the demure heroine--even including echoes of Balfe's famous 'I dreamt that I dwelt in marble halls,' from *The Bohemian Girl*. Doubtless it was partly to make the soprano's role more substantial that Loder shaped the solo into a scena, but the evident care with which he did so reveals his preference for dramatic situations. The recitative became an independent, carefully-shaped lyrical movement that has a Mozartian element, for the key (F minor), accompaniment, and idea of a fruitless search recall the beginning of the Act IV finale of *Le Nozze di Figaro*. It is especially impressive on account of its steadily-increasing intensity (6.17). The two verses are set as separate movements in accord with the usual scena design. The slow movement ('My heart is thine') bears a clear resemblance to Giselle's 'Ah no, you'll not forget me quite' in *The Night Dancers* (6.18) in its melodic shape (pivoting on the sixth degree) and rhythmic repetition, now applied with more sophistication by stressing the second beat of the bar. Though the parallel is not sufficiently close to be exemplified by specific passages, the allegro vivace is indebted for its quaver patterns to Agathe's 'All'meine Pulse' in *Der Freischütz* and the similar style of Reiza's 'Yes, my Lord! my joy, my blessing' in the Act I finale of *Oberon*. This is the weakest movement of the scena; the clash of textual and musical styles is unmistakable. The normal mood for such junctures is the heroine's eager anticipation of her beloved's company, but this sentiment is not stated sufficiently strongly to justify the joyous outburst. Moreover, Loder's small proportions prevent the musical ideas from being developed or given enough rhythmic variety.

The absence of a comparable piece for Raymond betrays Perren's indifferent ability. Whenever the tenor is called upon to sing at allegro tempo there is support from at least one other voice. The main solos in his part are two ballads and sections in the first of the two duets with Agnes and in the duet with Lindenberg. However, the tendency to set the voices in octaves in the love duets--a texture lacking in colour--is a weakness that must be attributed to Loder. Nor are the ballads fully satisfactory. The melodies lack the broad flow achieved by others such as Balfe. In 'Farewell the forest and the plain' the first phrase pauses in plainest ballad style at the end of textual lines, and thereafter the melody develops a polyglot flavour: the second strain has a Weberian clarinet-like quality, while the third is more Italianate (6.19).

The vocal score of the 1859 version includes a scena for Antoni at the beginning of Act III, presumably originally included at the request of the singer. It was omitted from the performances. The word-setting is among Loder's most careless--consider 'Shake to *its* deepest stone *yon* castle wall.'

A major difference from *The Night Dancers* is the extent to which Loder drew upon the style of Rossini and his successors, particularly in the extensive finales.[6] The trio 'Now in her eyes those trembling tears' in the first finale makes use of a characteristic dotted rhythm (6.20), and the finale of Act II, which was performed as a separate scene (the curtain fell and rose before it commenced) consists of the three phases standard for central finales, emphasized by the tonal scheme of E flat-B-E flat. First there is a phase of activity involving solos and chorus. The slow central section, 'Lost! and in a dream,' is an amalgam of Italian elements. It is in the canonic style popularized by Rossini; its use of the rhythm dotted-quaver semiquaver and triplet quavers recalls Bellini; and the effect of a high syncopated soprano over chromatic harmonies when approaching a climactic cadence bring to mind the central larghetto of the Act II finale in *Lucia di Lammermoor* (6.21). The final phase is marked by a conventionally energetic response to the situation. The concluding appearance of the 'St Agnes' theme may perhaps be attributed to Loder, for the words do not appear in the libretto.

The duets for Raymond and Agnes are cast in the sequence of cantabile and cabaletta movements established by Rossini, as in 'In these halls of ancient grandeur,' though the restricted vocal writing is distinctive. The first movement of 'Oh Agnes! cans't thou now forget' consists primarily of a solo for Raymond, with the usual individual response of the second voice omitted in consequence. To compensate (there is no textual reason) the soprano was given a long statement to open the second movement. The concluding joint section, primarily homophonic, is treated as a lengthy separate unit.

The duet for Raymond and the Baron is of an entirely different ilk. One of the most original and dramatic pieces in all English opera of this period, it is all the more impressive because the normal structure of melodramatic libretti does not allow weighty confrontations between hero and villain prior to the denouement. The only comparable cases are the duets for Podiebrad and Count Magnus in Wallace's *Matilda of Hungary* (1847) and for Raoul and Fabiani in *The Armourer of Nantes* by Balfe (1863). Both are much weaker. Loder probably proposed the design to Fitzball. The text incorporates more information and action than lyrical reflection. Rage and anger increase to the point where Lindenberg attempts to stab the hero, but Raymond grasps his dagger--only to see on it the name 'Inigo,' identifying the Baron as his father's murderer. The villain summons his men, and Raymond is thrown into a dungeon.

Loder divided the duet into three lengthy sections and a concluding tutti. First, in D major, comes a dialogue in which the characters are distinguished by melodic shapes and accompaniment figures, combined with harmonic effects. The second phase, Raymond's lyrical andante 'While yet in boyhood's rosy morn,' is highlighted by being set in B major. The graceful melody has a pentatonic inflexion that recalls the operatic use of folksong beloved of previous generations (6.22). It also has an Italian turn of cadential phrase: compare 'Alone in life, a child of grief!' with 'dell'amor tuo vivrò' in the love duet of Verdi's *I due Foscari* of 1844 (6.23). To some extent, however, Loder's piece is weakened by the strophic repetition of the same music for the different textual implications of the second verse.

For the final part the key returns to D major, providing a tonal pillar for the entire piece. Initiated by Raymond's cry for vengeance on his father's murderer (as yet unknown), it first develops tension by sequentially repeating the opening phrase. With the hero's insistent 'Agnes I demand to see once more' the voices are steadily brought closer in time, culminating in simultaneous singing. Raymond's seizure of the Baron's knife is marked by an unpredictable chromatic chord (F-sharp major, not the expected tonic) and a change of texture to recitative. There soon follows the large tutti passage that brings the dramatic episode to an end.

Formally the piece resembles a central finale, and in other ways it has more in common with a concerted piece than a normal duet. Indeed, certain details may have been suggested by the quartet in Act II of Beethoven's *Fidelio* ('Er sterbe'). There are similarities regarding the opening repetition of a stark phrase at the tonic and supertonic levels, the predominantly quaver rhythmic impulse, the variation of phrase lengths, and certain harmonic progressions. Such musical resources were in the public domain, but there is a definite affinity between Loder's duet--unique in English opera of the period--and Beethoven's ensemble. The quality of Loder's invention is further illustrated in two other regards. The heroine's prayer 'Oh Agnes! martyr fair' is unusual insofar as the accompaniment is not in the chromatic vein conventional for such devotions but maintains harmonic simplicity and a remarkable figuration throughout, scored with noteworthy sensitivity: violins maintain an inverted pedal over harmony that is sustained by bassoons, horns, viola, and double bass, with pizzicato cello (6.24). Another feature is Loder's arioso style, well exemplified by Agnes's plea for peace 'Men, who with relentless hearts' (6.25). The lyrical quality and richness of harmonic and instrumental texture are eloquent witnesses to the composer's sensibility.

SUMMARY

It is a sad reflection that only two operas give an adequate impression of Loder's talents. He was forced to spend the majority of his compositional life on hack work that allowed little chance for his creative abilities to flourish. No composer was immune from the pressures that forced this upon him, but in Loder's case it is a particular cause for regret. Though flawed--partly by circumstances beyond his control--*The Night Dancers* and *Raymond and Agnes* demonstrate that he was a composer of conscientiousness, skill, and imagination.

We should look to Giselle's death in *The Night Dancers* and the duet for hero and villain in *Raymond and Agnes* in order to appreciate

Loder's gifts. Neither piece has an equal in any other English opera of the time. Characteristically, he responded to a sequence of dramatic events, developing a structure of musical ideas whose growing intensity adds weight to the emotional situation.

Loder's professional life is a vivid reminder of the worst aspects of an English opera composer's lot. There is a strong sense of unfulfilled potential. English opera could ill afford the loss of a composer whose dramatic sensitivity and quality of musical invention were so unusual.

NOTES

1.Regarding the unsatisfactory circumstances of the performance, see Berlioz's letter to Loder, dated 15 April 1853, in Francis Hueffer, *Half a Century of Music in England* (London, 1889), pp.206-7.
2.This song is reproduced in *Musica Britannica* 43: *English Songs 1800-1860*, edited by Geoffrey Bush and Nicholas Temperley (London, 1979), pp.92-93.
3.*Athenaeum*, no.1651, 18 June 1859, p.816.
4.See the announcement in *The Times*, 3 June 1859, p.8.
5.*The Times*, 13 June 1859, pp.6, 9.
6.Nicholas Temperley suggested to me that a further sign of Rossini's influence is the similarity of the march in Act III (pp.223-28 of the vocal score) to the 2/4 allegro vivace section in the overture of *Guillaume Tell*, containing the well-known march theme. Both are in E and consist of eight-bar phrases, the first (which is repeated) starting with tonic harmony, the second with a chord of C sharp minor. It is difficult to be categorical about what might be common stock.

VII

Balfe

Michael Balfe (1808-1870) was the most prolific of all the composers of English opera during the period under review. Compared to Loder's two substantial contributions and Wallace's half-dozen, for instance, his total of twenty-one English operas, not to mention three French and four Italian works, is impressive. His first extended composition was a ballet, *La peyrouse*, given at Milan in 1825.[1] It was followed by the operas *I rivali di se stessi*, performed in Palermo in 1829, *Un avvertimento ai gelosi* at Pavia the next year, and *Enrico Quarto al passo della Marno* at Milan in 1833. The earliest of his English operas, *The Siege of Rochelle*, appeared in 1835, and his last completed operatic work, *The Sleeping Queen*, in 1864. Many of his operas were performed in various languages in major European cities. The period from April 1843 to November 1844 is unequalled. It saw two operas at the Opéra Comique in Paris, *Le puits d'amour* and *Les quatre fils Aymon* in April 1843 and July 1844, and two in London, one of which, *The Bohemian Girl* (November 1843), is among the most successful English operas of any century, while the other, *The Daughter of St Mark* (November 1844), was his second, and final, attempt to overcome the traditional English dislike of all-sung opera. The number of new works opening in London during these months rises to three if we add *Geraldine, or The Lover's Well*, a translation of *Le puits d'amour*. No other British composer was commissioned to write for two of the most prestigious opera houses in the world, Her Majesty's in London and the Opéra in Paris--the works being *Falstaff* (1838) and *L'Étoile de Séville* (1845). Balfe was fêted in France, Austria, and Russia. He was created a Knight of the Legion of Honour of France by Napoleon III, and a Commander of the Order of Charles III of Spain. The King of Prussia also wished to decorate him, but the offer had to be refused due to British protocol.[2]

Balfe's Career

Balfe's early teachers included William Rooke, best known for *Amilie, or The Love Test*, performed at Covent Garden in 1837, and C.F. Horn, a musician to the royal family. In his early days he was a violinist in the Drury Lane orchestra. His singing career began with a disastrous debut as Caspar in *Der Freischütz* at Norwich. In 1825 Balfe made his way to Rome and thence to Milan to study composition with Federici and singing with Filippo Galli, who created roles for Rossini and Donizetti. He moved to Paris and was introduced by Cherubini to Rossini. Returning to Italy, he made the acquaintance of Bellini ('vedendo Balff basso cantante salutamelo assai,' wrote the latter to a mutual friend in 1829[3]), and sang in various cities. His repertoire included a number of Rossini roles: Figaro in *Il Barbiere di Siviglia*, the Podestà in *La Gazza Ladra*, Dandini in *La Cenerentola*, and Iago in *Otello*, which he performed in May and October 1834 at La Scala, Milan, with Malibran as Desdemona.[4] (Iago's is a tenor role, but it was frequently adapted by transposition or puntatura.) On 14 August 1835 'Balfi' took part in a concert at the Vauxhall Gardens in London alongside the sisters Brambilla, Giulia Grisi, Kyntherland, De Angeli, Lablache, Ivanoff, Rubini, and Tamburini. This was exalted company: Grisi, Lablache, Rubini, and Tamburini were the famous '*Puritani* quartet.' The programme consisted of five items by Rossini, two by Bellini, and one by Donizetti.[5] In the late 1830s he sang in his own works as well as in operas by Barnett (*Fair Rosamond* and *Farinelli*), and in English versions of Bellini's *La Sonnambula*, Donizetti's *L'elisir d'amore*, Luigi Ricci's *Un'avventura di Scaramuccia*, Mozart's *Don Giovanni*, and the first English version of *Die Zauberflöte*.

As the following account indicates, until the early 1850s Balfe's fortunes were linked to Bunn's. Additionally he conducted at Her Majesty's from 1846 to 1852, with a period (1848-49) at the Theatre Royal, Manchester. During the lean years that followed Bunn's final downfall he travelled to St Petersburg, Vienna, and Trieste, resuming his operatic work in London when Pyne and Harrison formed their company in 1857. He retired to Hertfordshire in 1864.

Output

Balfe's composition of English operas falls into three phases, related to the circumstances of the London theatrical world. The first extends from 1835 to 1841, the period of Bunn's first management of Drury Lane. It saw *The Siege of Rochelle*, 29 October 1835; *The Maid of Artois*, 27 May 1836; *Catherine Grey*, 27 May 1837; *Joan of Arc*, 30 November 1837; *Diadesté, or The Veiled Lady*, 17 May 1838; and *Keolanthe, or The Unearthly Bride*, 8 March 1841. All were produced at Drury Lane except *Keolanthe*, which was given at the English Opera House, Lyceum, under Balfe's own management. These years also saw the incidental music to Planché's *Caractacus* (Drury Lane, 6 November 1837) and *Falstaff* (19 July 1838).

The second period begins in 1843 and lasts in effect until 1847, though two further works were produced in 1852. Largely it coincides with Bunn's second management of Drury Lane. Including Balfe's French operas, the works of this phase are: *Le puits d'amour*, 20 April 1843; *The Bohemian Girl*, 27 November 1843; *Les quatre fils Aymon*, 29 July 1844; *The Daughter of St Mark*, 27 November 1844; *The Enchantress*, 14 May 1845; *L'Étoile de Séville*, 17 December 1845; *The Bondman*, 11 December 1846; *The Maid of Honor*, 20 December 1847; *The Sicilian Bride*, 6 March 1852; and *The Devil's In It*, 26 July 1852. All the English operas were performed at Drury Lane, except *The Devil's In It*, which was given at the Surrey. (It was performed posthumously as *Letty, the Basket Maker*, in 1871). The first two Parisian operas were performed at the Opéra Comique, the third at the Opéra. At least two other works were left incomplete. *Elfrida* was abandoned because Grisi, for whom it was intended, became pregnant. Balfe also sketched parts of a setting of Victor Hugo's *Le Roi s'amuse*, begun at Manchester in September 1848. (The sketches are discussed below, in connection with *Satanella*.)

The final phase, lasting from 1857 until 1864, corresponds with Pyne and Harrison's management. Only 1859 did not see a new opera. The works are: *The Rose of Castille*, 29 October 1857; *Satanella, or The Power of Love*, 20 December 1858; *Bianca or the Bravo's Bride*, 6 December 1860; *The Puritan's Daughter*, 30 November 1861; *Blanche de Nevers*, 21 November 1862; *The Armourer of Nantes*, 12 February 1863; and an operetta, *The Sleeping Queen*, 31 August 1864. All were performed at Covent Garden except *The Rose of Castille* and

The Sleeping Queen, given at the Lyceum and the Gallery of Illustration respectively. *Pittore e Duca*, whose text was by Verdi's librettist, Piave, was performed in the interval between the second and third phases, at Trieste on 21 November 1854. It was given in London as *Moro, or The Painter of Antwerp* in 1882. At the time of his death in 1870, Balfe was working on *The Knight of the Leopard*, which was performed in 1874.

General Introduction to Balfe's Style

The salient features of Balfe's English operas are melodic fecundity and, in general, a preference for the idiom of Italian opera, primarily the buffa and semiseria types. His early command of the Italian style is shown by the aria 'Tutto parea sorridere' (by 'Signor Balfi'), published in London in 1835. Comprising an andante and allegro, it has various hallmarks of the type (7.1).[6] Balfe continued to write Italian arias, as substitute pieces, while conductor at Her Majesty's.

He avoided certain features cultivated by some of his English colleagues. There are few instances of the 'melodramatic' use of speech accompanied by music. Ensembles in the glee tradition popularized by Bishop are also rare; among the few examples are a 'Hunting Glee' and an unaccompanied quartet in Act III of *The Bondman*, and a 'madrigal' in Act I of *The Maid of Honor*.

Balfe's basic orchestra consists of double woodwind (including piccolo), four horns, two trumpets, three trombones, serpent (or serpentone or ophicleide), percussion, and strings. This is the same as in *L'Étoile de Séville*, written for the Paris Opéra. The orchestration is pragmatic without pretence to subtlety, occasionally crass. As a rule, upper woodwind and upper strings are used for melodic and upper-register orchestral passage-work, and the lower instruments, including brass, for bass and harmony. In ballads the harp frequently accompanies with arpeggios. The various obbligato solos, which are mostly for wind instruments, reflect contemporary usage. The popular cornet à pistons (used in *The Maid of Artois*, for instance) is an example. Doubtless the availability of suitable orchestral musicians influenced the unusual choice of obbligato viola in *Satanella* and the provision of an alternative clarinet part for the basset horn in 'The heart bow'd down' in *The Bohemian Girl*.

A particular influence on Balfe's operas was exerted by the tenor William Harrison. Harrison enjoyed an exclusive position at Drury Lane during the 1840s, and, like other composers, Balfe wrote his heroes' parts for him.[7] Harrison was not alone among tenors in preferring flat keys--A flat, D flat--for slow movements, but the frequency of this choice, his upper limit of B flat, and his lack of flexibility, especially later in his career, make such points characteristic features.

Balfe worked at a tremendous pace. Two days after the premiere of *The Daughter of St Mark* he wrote to St Georges that

> We succeeded in giving our premiere of "La fille de St Mark" on 27 November--anniversary of the Gipsy--Its wonderful I received the first lines of this work on 28 Sept 2 months and 7 days to *compose*, *rehearse*, and *perform* a grand opera in 4 acts![8]

A memorandum of December 1844 indicates that he allowed three months to compose an opera.[9] Dates in his autograph score of *The Rose of Castille* show that its composition took from 19 September to 11 October 1857, leaving a few weeks for rehearsals before the first night on 29 October. In this case the time required was reduced by borrowing previously-composed material, but such swiftness points to considerable facility. It is also related to a lack of dramatic depth, for rarely is speed conducive to reflection.

In the ensuing discussion individual operas have been selected to illustrate Balfe's work in the three phases of his career, placing these particular exemplars in their context by means of less detailed accounts of other operas.

THE EARLY PHASE

The Siege of Rochelle

The extraordinary success of *The Siege of Rochelle* established Balfe's fame overnight. Fitzball, the librettist, recalled that 'It was a glorious night, the first night of "The Siege of Rochelle"--one to wish your whole life long the first night of a new play or a new opera ... never had I witnessed anything like it.'[10]

It was the most Italianate of the English operas of its day. Indeed, Fitzball's remark that 'in a *day* or *two* the first act was complete'[11] suggests that Balfe may have used material written in Italy. The familiar two-act structure was also used for Luigi Ricci's *Chiara di Rosemberg* (Milan, 1831), a setting of the same plot, in which the heroine, Clara, is unjustly accused of murdering the child of her beloved, the widower Valmour. With the aid of the honest Michel, her innocence and the guilt of Montalban, Clara's supposed father, are revealed. (Ricci's second opera derived from Mme de Genlis's novel, *Chiara di Montalbano*, was first performed on 15 August 1835, the day after Balfe took part in the Vauxhall concert.) Balfe's overture, cast in two movements (slow--fast, moving from major to minor mode) is clearly modelled on Rossini's style. The parallel is particularly clear in the second subject of the allegro. This is introduced by a characteristic extended dominant and opening bar of tonic harmony, with a steady tramp in the accompaniment. The melody itself is virtually a hybrid of the second subject of *La Gazza Ladra*'s overture and the march in Act I of *Zelmira* (7.2). Certain rhythms add to the overture's Italian flavour: the Bellinian dotted-quaver-semiquaver followed by triplet quavers in the opening strains, and minim tied to the first of four quavers, minim syncopated across the middle of a bar, and syncopated crotchets in the allegro. But Balfe's work is not mere unimaginative imitation: the syncopated crotchets are preceded by an effective harmonic digression (7.3). On the other hand, the sixteen bars that emphasize the second subject's key (bar 120ff.) are too obviously padding.

In writing Michel's buffo aria in Act I, 'Trav'llers all of ev'ry station' (which he himself sometimes sang[12]), Balfe must have had Figaro's cavatina in *Il Barbiere* and Ricci's equivalent in *Chiara di Rosemberg* in mind. Similarities of text put his knowledge of the latter beyond doubt; Fitzball's jingoistic comments about England replace Rossi's chauvinistic references to Milan. Both Balfe's and Ricci's settings are lengthy solos with occasional choral interjections, but Balfe followed Rossini's example more closely than did Ricci. This is most evident in his 6/8 allegro vivace section--the same metre and tempo as Figaro's. It is too short to accommodate the variety of ideas included in 'Largo al factotum,' and to some extent compensates for this with a variety of tonal colours, ranging several degrees sharpwards (A and D in the midst of B flat). The sudden move to ♭VI brings Rossini's technique forcefully to mind--compare 'England's the garden of lovely women' with the approach to 'Rasori e pettini' (7.4).

Valmour's cavatina 'In vain my soul her image flies' is the only aria to be preceded by recitative. Balfe conceded to the English distrust of recitative by writing this in an unusual style that verges on arioso (7.5). (The sole other instance of recitative occurs at the beginning of the trio in Act II.) The hero's two-movement aria conforms more closely to the Italian mode. The larghetto includes a 12/8 version of the dotted quaver-semiquaver, triplet rhythm noted earlier, coupled with a typical 4-3 suspension. The allegro draws on the repertory of the Italian heroic tenor, with arching contours and familiar dotted patterns and syncopated minims in the vocal line.

Only one Rossinian rhythm is noticeably absent: semiquaver-dotted quaver-semiquaver. Apart from this, examples of the Italian atmosphere could be multiplied--for instance, by details in the duet section of the first quartet ('Sweet blossom of paternal joy'), and various orchestral figurations (compare the pattern at Michel's entry in the Act II trio with Mathilde's 'Pour notre amour' at the beginning of Act III of *Guillaume Tell*). In one instance the specific example of Donizetti's *L'elisir d'amore* arises. Michel's Romance, 'When I beheld the anchor weigh'd' was clearly inspired by Adina's 'Prendi, per me sei libero' (7.6). The similarities include the opening basic melodic shape, the accompaniment pattern, the descending orchestral scales in thirds, and the location of cadential flourishes. We may also note the shared use of falling fifths in the waltz movement ('Thus together, side by side') of Michel and Marcella's duet and in the 3/8 allegro vivace movement ('Così chiaro è come il sole') of Dulcamara's cavatina in *L'elisir d'amore* (7.7).

The first phrase of Valmour's larghetto concludes with major mediant harmony, a common effect taken one stage further in Clara's ''Twas in that garden beautiful,' where the mediant is followed by a dominant seventh in the tonic. Essentially a reharmonization of the leading note, this elliptical progression is a harmonic cliché of the period: to quote only a selection of possible examples, it occurs in Bellini's *La Sonnambula*, Donizetti's *Lucia di Lammermoor*--first performed a month before *The Siege of Rochelle*--Meyerbeer's *Les Huguenots*, Verdi's *Macbeth*, and Wagner's *Lohengrin*. Balfe used it in his French and English works--as witness the occurrences in Estrelle's romance 'Sais-tu bien quel tourment' in *L'Étoile de Séville*, and 'I dreamt that I dwelt in marble halls' in *The Bohemian Girl*.

There is no equivalent in Ricci's setting for the choral prayer in the Act I finale (7.8). The best-known precedent for a prayer in this context was 'Dieu puissant que j'implore' in the Act III finale of Auber's *Fra Diavolo*, one of the most popular of contemporary French operas, though in its harmonic richness Balfe's is more reminiscent of another predecessor, 'Ciel qui du monde est la parure' in *Guillaume Tell*. (The latter is a separate unit, not inserted in a finale.) However, Balfe may well have been forced to make a virtue of necessity. At the heart of the issue was the solo soprano. Ricci's alternative to the prayer in his finale was a second solo passage for the prima donna, Giuditta Grisi, sister of the more famous Giulia. Contrasted with her powers, those of Balfe's soprano, Miss Shirreff, must have been very limited. ''Twas in the garden beautiful,' largely restricted to crotchet movement, confirms this. Indeed, with the exception of the roles Balfe intended for specific virtuoso singers, all his English operas are marked by vocal writing that is much less virtuosic than its Italian equivalent. Lack of vocal stamina may also explain the comparatively small number of arias and the compensating number of ensembles, which include two quartets in Act I and three duets in Act II. The low level of expertise among singers taking smaller roles influenced the simple writing in these ensembles. The canonic quartet, 'Lo!, the early beam of morning,' is a case in point. (Its rhythm, in effect a dotted crotchet and five quavers, appears frequently in the opera.) The villain has a minimal vocal role.

The ballad, soon to become an essential component of Balfe's operas, is absent at this stage. The closest approach is Clara's Romance ''Mid the scenes of early youth' (7.9). It does not have the broad melodic line typical of the genre, and its chromatic part-writing and harmonic progressions, notably at the end of each verse, are also unlike the ballad's more chaste style.[13]

Other Operas of Balfe's First Period

Balfe's next opera, *The Maid of Artois*, presented an opportunity to develop his style. It was the first of his operas conceived for Drury Lane (*The Siege of Rochelle* had been originally intended for Arnold's English Opera House), and the first written for a virtuoso singer--the beloved mezzo-soprano, Malibran.

The Italian element is very strong in her role. Set predominantly in D and E majors--the main keys of Rosina's mezzo role in *Il Barbiere di Siviglia*--her solos are replete with typical vocal shapes and accompaniment patterns, for instance in the Act I duet 'Oh! feel for one' and the solo section 'Here take my life' in the Act I finale (7.10). The exception arises in the ballads. It is in this opera that Balfe's ballad style, which was to stand him in good stead for over a quarter of a century, makes its first appearance. There is one example in each act. Soon to become one of Balfe's most celebrated ballads, 'The light of other days' takes the place of what might have been an Italianate solo in the Act II finale (7.11). We will investigate the ballad further in discussing *The Bohemian Girl*.

Malibran died in September 1836, well before Balfe's next opera, *Catherine Grey*, received its premiere (May, 1837). This is more eclectic: the overture includes allegro D major semiquaver arpeggios reminiscent of Weber, a rare instance of similarity to the German composer in Balfe's work. Some of the song-writing is blatantly commercial. But in other regards the opera is strongly based on the Italian model. The subject-matter is of the same ilk as Rossini's and Donizetti's 'Tudor' operas (the former's *Elisabetta, regina d'Inghilterra* immediately comes to mind), and *Catherine Grey* approaches them more closely by being all-sung. The recitative uses orchestral figures that vary in familiar fashion from rhythmic gestures to tremolo accompaniments. The two female characters, Catherine and Elizabeth, each have a 'grand scena,' consisting of the conventional two movements with preceding recitatives and intervening episodes, as is found plentifully in the works of Bellini and Donizetti. During her scena Elizabeth has a parlante passage set to a rigid rhythm, largely on one note, and accompanied by woodwind chords. Balfe marked the words to be 'almost spoken'--an unsubtle conception of theatrical speech (7.12). Here, as one short section among others, the passage adds variety, but this method of setting English poetry is potentially wearisome. Elizabeth's cabaletta, 'Hence! thou base, perfidious slave' (7.13), is particularly Italianate, with its vocal line that progresses towards familiar high syncopated crotchets and fast semiquaver scales, and a polonaise rhythm in the orchestra.

The *Athenaeum*'s critic described *Catherine Grey* as 'by no means so good as 'The Siege of Rochelle,' but much less dull than 'The Maid of Artois.'' A noisy claque employed by the publishers could not engender much acclaim. 'The only thing to bring away with one, was the serenade ... sung, and very charmingly sung, by Mr. Balfe himself ... [with an] exaggerated Tamburini cadence [cadenza].' Miss Romer (Elizabeth I) had already 'expressed a strong wish to give up the part.' As regards its use of recitative and the quality of the libretto, the writer opined that 'we do not think the English language well fitted for this purpose, but we are quite sure that *such* English language is not.... The plot, too, is not only poor, but absurd.'[14] Balfe now set his face against the Italian idiom in an effort to regain popularity. Rossinian traits are noticeably absent from *Diadesté*, a small-scale comic opera characterized by minimal recitative and limited vocal writing, while *Joan of Arc*'s choral scenes recall the imposing style of Auber's and Meyerbeer's grand operas. The solo parts in this opera are remarkably constricted, a feature all the more noticeable because the heroine was played by the same soprano as had sung Elizabeth I in *Catherine Grey*, Miss Romer. (Balfe took another solo part.) The solos seem primarily intended for the commercial market. Indeed, the entire opera gives the impression of having been cobbled together without regard to dramatic effect. The playbill for the first performance explained that 'to heighten some of the effects, a new & extraordinary light will be introduced, called PHOSHELIOULAMPROTERON.' Whatever this was, it could not add brilliance to the opera, which was replaced by *La muette de Portici* within a month. *Joan of Arc* was dedicated to the newly-crowned Queen Victoria; not a pretty dish to set beside Prince Albert. The most one can say is that Fitzball's libretto got what it deserved. Balfe probably devoted much more thought to *Falstaff*. However, this turned out to be an unexceptional opera buffa, and, despite the advantages of being performed by the '*Puritani* quartet,' it did not hold a place in the repertoire.[15] The few remaining sources of *Keolanthe* suggest that it was uninspired. Balfe's wife took part. She had been a professional singer, but one senses an element of *faute de mieux*.[16]

BALFE'S SECOND PHASE

The Bohemian Girl and *The Daughter of St Mark*

After the collapse of his managerial hopes at the Lyceum in the spring of 1841, Balfe decided to develop his career in Paris. In August of that year *La France Musicale* announced that 'Balfe, l'Auber de l'Angleterre,' had arrived in Paris to work on an opera for the Théâtre Italien, to be sung by Grisi, Mario, and Tamburini.[17] This was *Elfrida*, brought to naught by Grisi's pregnancy. *Le puits d'amour* was produced at the Opéra Comique in April 1843, seven months before *The Bohemian Girl*, and a further French opera appeared during each of the next two years. By this time, Balfe was working concurrently in Paris and London.

It is not surprising, therefore, that one cannot speak of a single style during his second period of English operas. The use of foreign idioms fluctuates. *The Bohemian Girl* accommodates a mixture of styles. (The opera had been proposed by Bunn some years earlier; some of the composer's initial work on the piece may well have been used in intervening projects.) To place it in context, we may note that the first season of Bunn's second administration of Drury Lane opened on 30 September 1843 with *The Siege of Rochelle*, and continued with Bishop's *Clari* and English versions of Weber's *Der Freischütz*, Auber's *Fra Diavolo* and *Gustavus III*, Donizetti's *L'elisir d'amore* and *La favorite*, and Rossini's *La donna del lago* and *La Cenerentola*. Balfe directed the first eight nights of his new opera (Benedict resumed the conducting thereafter). The eighth was Balfe's benefit night. After the opera, the usual ballet--*The Devil in Love*, on which *Satanella*, of 1858, would be based--was replaced by a concert. Balfe and his wife sang two of his own arias and two duets by Donizetti, and Benedict and Madame Dulcken played Thalberg's 'Gran Duetto Concertante' on themes from *Norma*. Then came the pas de trois from *La favorite* and, finally, a farce. Only two days later the Balfes left London for Paris.[18]

When planning *The Bohemian Girl*, probably Balfe and Bunn decided jointly to have the minimum of recitative in the search for an operatic formula that would appeal to audiences. In this regard, the opera returned to the style of *The Siege of Rochelle*. Balfe also revived some of the Italian element that he had reduced in more recent works. Having discussed the use of such traits in *The Siege of Rochelle* in some detail, we need not consider them as fully here. They range from the semiquaver figuration of the overture's second subject--lending itself to a crescendo--to Figaro-inspired buffo writing for Florestein and the use in the Introduction of Act II of a rhythm similar to the march in *Zelmira* (as in *The Siege of Rochelle*). From Rossini's example, too, come the idioms of the Act II duet for tenor and soprano and the canonic texture of the Act III quintet. The rhythm of dotted quaver-semiquaver, triplet, is often prominent. The recurrent melodramatic music is of the same stock as the D minor allegro assai section of the Act I finale of *La Sonnambula* (7.14). There are also some French elements. *Guillaume Tell* probably supplied the model for the chromatic basses in the overture, and for the gypsy chorus--compare the Tyrolian chorus (7.15). The Austrian Soldiers' march in Act I recalls the overture to *Guillaume Tell* as well as another of the latter's progenies, the 2/4 allegro section in the ballet beginning Act V of *Les Huguenots*. The ballad 'I dreamt that I dwelt in marble halls' calls upon two features associated with Auber, the partiality for a slow harmonic pace and for pedal basses. These are linked to a melodic shape that spreads the tonic triad over a broad span.

The overture is unusual in Balfe's output in that he sought out musical material from the world of the plot. He probably believed he had based the first eighteen bars of the allegro on a fifteenth-century Hussite melody made popular by Liszt. Indeed, Balfe's first six bars are virtually identical to the 'Version litterale' printed with Liszt's piano piece based on the melody, published in 1840 (7.16). In fact the song was written by one Joseph Kroff [Krov] in 1830. The idea of adding colour in this way would have been suggested by Rossini's incorporation of characteristic melodies in the overture to *Guillaume Tell* and, more specifically, Meyerbeer's use of 'Ein feste Burg' in *Les Huguenots*. The French operatic element also comes to the fore in the use of the waltz and the galop. These dances had captured widespread enthusiasm, but, in particular, Auber's inclusion of a galop in *Gustavus III* was doubtless the spur for its appearance here, as a ballet movement in Act I and as the background to Arline's monotonic 'Sir knight and lady, listen!' in Act II. Similarly, the dance rhythms underlying the introductory chorus of *Fra Diavolo* find an echo in the chorus that opens the Act III finale. On the other hand, the gypsy queen's warning to the Count in the Act III finale--intoned on a monotone over lugubrious chords--is in a tradition of cataclysmic statements that includes the Commendatore's in *Don Giovanni*.

The Ballad

The Bohemian Girl is particularly known for its ballads, of which there are three--one each for soprano, tenor, and baritone--and we should now turn attention to this important element of English opera.

In composing ballads, Balfe responded unerringly to contemporary taste. Apart from those included in operas, he also wrote many for the domestic market. These tend to be less demanding for singer and accompanist (see, for example, 'A home in the heart'). The following discussion concerns his operatic ballads.

Historically, they are descendants of Bishop's songs--'Home, sweet home' in *Clari*, or 'The sun is o'er the Mountain' and 'In my bower a Lady weeps' in *Aladdin*, for instance. As has been pointed out, Thomas Moore's example lies behind the genre. A restrained mood is characteristic. The key is always major, usually in flats (E flat, A flat, D flat--allowing a top B flat in the case of tenors, and suitable for the harp). Typically, the pulse is a moderate tempo in 4/4, whereas Bishop was more varied in his selection of metres. The melodies move with cultivated simplicity, often in an arch-like shape, and are longer than Bishop's. Appoggiaturas are frequent. The phrase structure is uncomplicated, usually in ABA form. Its basis is a four-bar period ('The light of other days' from *The Maid of Artois*, for example, is cast in three units, 4 x 2, 4, and 8 bars), though the final phrase may be varied, and often includes an operatic flourish--another difference from Bishop's style. The harmony is sometimes more impassioned at such points, but otherwise is slow-moving and predominantly diatonic. The accompaniment is equally simple. A favourite pattern is quaver arpeggios, often scored for harp, over a slow-moving bass. Such orchestration was not exclusive to the ballad: the scoring for 'Torn from all I love' in *Catherine Grey* (melody--flute and clarinets; arpeggios--harp; harmony--horns, bassoon, and pizzicato strings) is also used for Estrelle's Act II romance 'Sais-tu bien quel tourment' in *L'Étoile de Séville*.

Written for Harrison, 'When other lips' (7.17.a) is probably Balfe's most famous tenor ballad. The song swiftly gained enormous popularity, selling over 80,000 copies in its first year. Harrison's son reported the difficulty Balfe had in reaching this deceptively simple setting,[19] as is confirmed by two sketches, BL, Add.32669, f.9v (7.17.b) and Add.32672, f.33 (7.17.c), found in sketches bound into books. These cannot be dated more precisely than 1836-43 and 1843-61, and it is not possible to gauge which sketch is earlier.[20] Only a portion of the words, and enough of the harmony to act as an aide-memoire, were notated. The key was altered--Add.32669 is in D flat, Add.32672 in A flat--and the phrase lengths, rhythms, and melodic shapes were reworked. Add.32669 is in four regular four-bar phrases, Add.32672 in five. The rhythm was troublesome. Balfe experimented with the placing of the quicker quaver movement in the first or second half of the bar. Add.32669 has a less happy melodic shape; the octave and sevenths give a more strident, slightly martial feeling. Add.32672 is smoother, and emphasizes the high A flat in the last pre-cadential bars, though this pitch has already been used.

Doubtless there were further sketches. The final version, in D flat, is more sophisticated than its predecessors. One barely notices that it starts with dominant harmony or that the rhythm of each phrase is subtly different, and that Balfe now emphasized bar-lengths, not half-bars--promoting the genre's characteristic sense of reflection, untrammelled by obtrusive rhythmic emphases. The melodic contour has flowing archlike shapes and carefully-weighted points of tension. The mediant cadence was among the later thoughts.

The penetration of Italian traits is particularly interesting. The dotted-quaver, semiquaver, triplet rhythm is notated in standard form in the orchestral introduction, but in a freer version in the second bar of the vocal part--the figure is, after all, no more than a notated expressive device, almost improvisational rubato. Balfe would have heard the effect of a climactic diminished seventh in approaching the final cadence, with the voice held aloft on a high pitch, in Amina's 'Come per me sereno' in *La Sonnambula* (7.17.d). He extended the final phrase to six bars by means of this very progression. Apart from these shared elements, a comparison of Balfe's and Bellini's pieces reveals the distinctive essence of the ballad--primarily its constrained melodic shape and rhythmic vocabulary.

No sketches for 'I dreamt that I dwelt in marble halls' (7.18) are known. The only light to be shed on the compositional process is that Balfe made some alterations to the text.[21] Probably the musical phrasing suggested inserting 'the' before 'pride', and reducing 'besought' to 'sought'. Strangely, a comparable replacement, 'neath,' was changed back to 'within' in the vocal score. Other changes--'dreamt' for the more poetic 'dream'd', 'But' for 'And' and 'Yet', 'that' for 'this' (verse 2, line 5)--may have been due to personal preferences regarding syllables or to unconscious alterations during composition, not re-altered later. Writing 'pleased' (changed to 'pleas'd' in the vocal score) in place of 'charm'd' in verse 1, but leaving the original word in verse 2, could be attributed to the same reason, perhaps followed by a liking for the resulting variety.

The scoring is based on Balfe's usual choice for ballads: the melody is delivered by flute, clarinets, muted violins, and cello, the semiquaver arpeggios by harp and viola, the off-beat harmony by horns and bassoon, the on-beat harmony by oboe, horn, and pizzicato basses. Its 3/8 metre, however, is exceptional, and brings it closer to an Italian romanza (for instance, Donizetti's 'Una furtiva lagrima' in *L'elisir d'amore* is in 6/8). Indeed, in the two later Italian versions of the opera, each titled *La Zingara*, it is called 'romanza'--'In una reggia splendida' in Paderni's translation for Trieste (1854) and 'Sognai che fra catene' in Zaffira's version, performed at the Théâtre Lyrique, Paris, in 1869--and the reduced number of syllables of the settenarii lines allows the rhythm to be smoother (see 7.18.b, taken from the latter translation). The melody's skeleton is the tonic triad, enhanced by cellular repetitions that incorporate a counterpointing of the basic pulse and the vocal emphases within each bar. These emphases stress a constant pitch on the downbeat and a higher pitch on the second quaver. The two emphases shift subtly, overlapping and then again separating. The repetitive units of the voice's initial phrase, over the prolonged tonic harmony, remove a sense of pulse and time. The harmonic pattern is Rossinian: two bars of tonic, one of dominant, and one of tonic, with a prominent tonic pedal. The pulse is quickened only in approaching the mediant cadence, where the delicate chromaticism adds a new shaft of harmonic colour. The cadence itself owes much of its charm to leaving the leading note unresolved, an effect that is immediately followed by the unexpected scoring of the first phrase which now returns--the orchestra doubles the vocal melody for the first time. The extension of the final phrase is brought about without disturbing the diatonic quality of the harmony and the dynamic, for the added propulsion stems mainly from the voice's ascending scales. One is never quite sure whether there will be a further scale before the final note is reached, such is the cumulative effect, almost mesmeric, of all the preceding repetition of melody, the lulling pulse, and the static harmony.

Two other sketches, Add.32669, f.28 (7.19.b) and f.51 (7.19.c), concern Thaddeus's aria ''Tis sad to leave our fatherland' (for the final version, see 7.19.a). Both settings are prosaic, except for the long cadenza in f.28's setting. Again, the key was open to change: f.28 was written in D, but Balfe later added 'in E♭.' Presumably this was to take advantage of the more resonant vocal quality gained by the higher tessitura. (The version on f.51 is in E.) Word-setting was a subservient consideration; melody was the priority. Balfe's autograph manuscript of the full score implies that the final version of this piece was one of several items decided upon at a late stage. At the end of the preceding chorus there is a note 'Segue Melo[..?] & new Song.' Then comes the melodramatic music, on a sheet of paper inserted separately, but there is no setting of ''Tis sad to leave our fatherland,' nor of its foregoing recitative; the next item is the gypsy chorus.

The Bohemian Girl is a tauter work than its predecessors. The repetition of material--the gypsy chorus, 'The Dream,' and the recollection of Thaddeus's account of Arline's history--has a valid dramatic point; even the recurrent instrumental melodramatic music, while open to accusations of expediency, adds appropriate effect. (Such instrumental sections were required by the amount of stage movement in this adaptation of a ballet.) The structure of the Act II Introduction shows a firmer hand, being unified by a recurring phrase. There is greater concern for textures, as in the counterpointing of the melody of the solo quartet 'From the valleys and hills' with the ensuing gypsy chorus. The wider emotional range of the hero's cavatina ''Tis sad to leave our fatherland' is noteworthy. Its opening section colours the allusion to emotional isolation ('orphans of the heart') with ♭III chromaticism, and the faster central section in the ABA-coda design is made more highly charged by a less melodic line and an orchestral figuration that, while perhaps suggested by the chromatic writing and scales in Rossini's duet for Otello and Desdemona in *Otello*, makes effective play on an enharmonic ♭VI effect (7.20).

However, we should not ignore the opera's flaws. The stretta of the love duet is dull, its melody wanting variety, and the conclusion of the Act II finale lacks the breadth and rhythmic drive achieved by the Italian masters. The first phrase of Thaddeus's aria 'When the fair land of Poland' is an example of the poor word-setting that mars the work. The verbal sense is left hanging in mid-air (7.21). Sims Reeves, self-preening to the end, reported that he once tried to make the sense clearer, but Balfe would not have it.[22] Moreover, two features of the work mask weaknesses that will emerge in our discussion of *The Daughter of St Mark*. It sidesteps the issue of effective setting of recitative by the expedient of having virtually none; and the libretto did not present the challenge of having to invent extended music for that essential melodramatic character, the villain.

The Daughter of St Mark took English opera closer to the world of the Paris Opéra. The action is set in fifteenth-century Venice. The marriage of Catarina and Adolphe is interrupted by her uncle, Andréa, who, under pressure from Moncenigo, a member of the feared Council of Ten, has agreed that Catarina shall marry Lusignano, King of Cyprus. Catarina is forced to deny her love for Adolphe in order to save his life. Later, at the preparations for the royal wedding in Cyprus, the lovers meet by chance and Catarina reveals the truth. They are imprisoned as traitors. At Catarina's trial, Andréa reveals that she is actually Moncenigo's daughter. The latter, stunned, now proclaims her innocence and his own guilt. To general rejoicing, the magnanimous Lusignano relinquishes his claim on Catarina's hand to Adolphe. St George's *La Reine de Chypre*, cast in five, not three, acts, ends differently. Catarina and Lusignano marry, but the king is slowly poisoned by Moncenigo.[23] Adolphe returns to inform Lusignano of the villain's evil act. Lusignano's death leaves Catarina Queen of Cyprus, and, with Adolphe at her side, she rejoices in the victory of her troops over the Venetians.

Part of the weakness of *The Daughter of St Mark* stems from Bunn's adaptation of St Georges's libretto. 'We doubt whether Folly could further go,' commented Chorley in his review.[24] Bunn's text corresponds up to Act III of the original. Thus it includes an aria for Catarina in Act II, but excludes the aria for Adolphe in Act IV of the French version. To compensate, Bunn gave Adolphe a ballad in Act III, and another for Andréa in Act II, even though the latter disrupts the dramatic action badly. He also supplied an aria for Moncenigo in Act II for which, again, there is no equivalent in the original, nor in Donizetti's version, *Catarina Cornaro*.

Balfe's music incorporates too wide a range of national idioms to be consistent as an entity. Only a few days earlier Chorley had remarked on Balfe's 'careless and unscrupulous triviality, equalling that of the most loudly reproached foreign tune-spinner, but unaccompanied by the nationality--or call it style--which redeems the Donizettis and

Adams from utter insipidity. Mr. Balfe's music is neither wholly Italian, German, French, nor Irish.'[25] *The Daughter of St Mark* makes far less use of Italian idioms than *The Bohemian Girl*. The composer's increased involvement with French opera was responsible. The inclusion of ballet movements is an obvious point of general similarity with Parisian models, but Balfe also followed Halévy's setting of *La Reine de Chypre* in specific ways. The duets in Act I for Adolphe and Catarina and in Act III for Adolphe and Lusignano were originally identical in design to their French equivalents (some material was later deleted). Certain other details are very close, as witness the falling melodic figures at 'Doux instants!' and 'Thou art mine' in the Act I duets (7.22), and the key schemes in the Act III duets--Halévy's has the progression G| A flat-B| G, Balfe's the enharmonically identical A| B flat-D flat| A.

Harrison, Borrani, and Miss Rainforth (Adolphe, Andréa, and Catarina) had played the major roles in *The Bohemian Girl*, each having a treasure-trove of a ballad.[26] *The Times*'s reviewer correctly predicted that no solo item in the new opera would achieve comparable popularity. The men's solos were well received--thanks to 'Oh, smile as thou wert wont to smile' Harrison was guaranteed acclaim without the extra work of having to sing a taxing aria--but evidently the critic was less impressed by the heroine's ballad, 'The Gondolier.'[27] Balfe made no attempt to copy 'The Dream' when writing the latter, which was placed immediately before Catarina's scena at the beginning of Act II. Set in the usual 4/4, its leaps and fioriture are slightly wider and more intricate than normal, but in other regards it is unexceptional. Regarding the scena itself, the French version consists an andante, equivalent to 'The Gondolier,' followed by an interlude in which Catarina discovers a message intimating that Adolphe will shortly come to take her away, and then an allegro, a joyful prayer. Only in its musical details does this format differ from the Italian format of slow movement followed by fast cabaletta. Having used 'The Gondolier' as Miss Rainforth's ballad, Bunn now produced a two-movement sequence for the soprano by using the prayer as the slow movement of the scena, penning the lines of a new allegro expressing eager anticipation, and repositioning episodic material.

Apart from the richly-harmonized prayer, Balfe's setting of the scena is unconvincing. He followed Halévy's notion of orchestrally-accompanied declamation for the episodic sections, but whereas the Frenchman's vocal line and accompaniment are well varied, Balfe was too content with unrelieved monotonic delivery and unvaried figurations. This was the style he had used to a limited extent in *Catherine Grey*. Here, when Catarina 'reads' Adolphe's letter aloud, 'Quasi Parlate,' it tiresomely emphasizes the tramp of the verbal stresses (7.23). (Donizetti's soprano simply scans the letter quickly.) The degree to which Balfe had forsaken the Italian style is especially clear in the concluding allegro, more so when we recall Elizabeth's scena in *Catherine Grey*. The vocal and orchestral material is unlike either an Italian or a French equivalent. The setting reminded Chorley of Weber ('the well-worn *scena* of prayer, expectation, and rapture, imitated from 'Der Freischütz''[28]) and the key, harmonic style, and certain vocal phrases of the last movement do indeed bring to mind Agathe's allegro--and Leonora's in *Fidelio*. It is the closest Balfe came to the German style in his vocal music. Perhaps Miss Rainforth requested quaver scales and top Bs in the manner of these heroines. Whatever the reason for its Germanic flavour, however, the movement compares badly with Weber's and Beethoven's efforts. It is too short--so much so that the melody cannot develop (7.24). As a whole, the scena accommodates such an uneasy mixture of styles that it has no internal consistency.

The other major aria, a single movement for Moncenigo, was sung by Balfe's pupil Weiss, and could well have been included as a result of personal pressure. It illustrates Balfe's limitations in inventing dramatic music suitable for a villain. Moncenigo's text has been discussed in chapter 2. The orchestral introduction is trite. Its triplet figure and contrast of an octave-scored dotted pattern with fuller harmony, also used to introduce the preceding recitative, were probably suggested by Halévy's motifs at Moncenigo's entrance (7.25). The tonal ambiguity of major/relative minor may be traced to the same example. Balfe's ideas do not have the range, variety, and bite necessary to characterize this malevolent man. The variations in tonality and the syncopation that come later in the aria cannot compensate for the lack of incisive melodic lines and dramatic harmonic effects.

Balfe's ineffective handling of heroic couplets is revealed all too starkly in Moncenigo's 'Cyprus has chosen from her shores to chace' (7.26). Halévy's equivalent is 'De Chypre le peuple coupable,' for which he devised a melodic line of rising tension that Balfe used in the Act III dialogue for hero and heroine. Balfe chose a parlante texture dominated by vocal monotones. The heavy accentuation of the iambics is made thuddingly worse by the composer's insensitive word-setting: consider '*of* her *king*-ly *race*' and '*of* al-*li-ance*.' The end of the preceding recitative is nonsensical (7.27).

Though much less in evidence, Italian traits are not, however, entirely absent from the opera. Andréa's ballad includes a turn of phrase very similar to one in Rosina's 'Io sono docile' in *Il Barbiere*: compare the harmony and 4-3 suspension at 'the bitter world neglecteth' with those at 'e cento trappole prima di cedere.' The most sustained use of Italian procedures occurs at the ends of Acts I and III. The larghetto of the Act I finale inevitably brings to mind the comparable interruption of a marriage in the Act II finale of *Lucia di Lammermoor*--in the same key, D flat (7.28). But Balfe's melodic talent flagged here; moreover, an effect such as the octave-scored forte upbeat that introduces the reprise merely emphasizes the dubious word-setting and is harmonically bald. Likewise, the Rossinian trio of astonishment in Act III ('If true his tale') is weak in invention.

The Daughter of St Mark is disappointing. The circumstances of Balfe's professional life were partly to blame. We may refer to his letter to St Georges in which he described the speed of composition. In the same letter he speaks of being with St Georges in Paris in a few days' time--recalling his movements shortly after the premiere of *The Bohemian Girl*. Clearly he was shuttling to and fro between the cities. This affected even his forte, melodic inspiration. Bunn's text, too, was an impediment. An unusually large amount of the original libretto was omitted. But the crucial weaknesses of *The Daughter of St Mark* are endemic to Balfe's musical personality. Too often he simply did not care enough. He was complacent about inappropriate dramatic construction, especially where the ballad was concerned. There is a cavalier attitude to word-setting, including in the recitative, which, in order to gain support for all-sung opera, should have received extra attention.[29] The limitations of his imagination are especially evident in terms of characterization. Finally, the opera has no stylistic consistency.

Other Operas of Balfe's Second Period

The remaining operas of this period have various individual features. *The Enchantress* is marked by its virtuosic writing for Thillon, for whom Balfe had already written *Le puits d'amour*; surprisingly, her virtuosic aria 'Who has not heard' includes 'melodramatic' spoken delivery. *The Bondman* and *The Maid of Honor* are uninspired pieces. Sims Reeves sang in the latter and in *The Sicilian Bride*. Perhaps to demonstrate Reeves's versatility, Balfe varied the ballads compared to the model favoured by Harrison, with a particularly chaste melodic line and unusual accompaniment figuration in 'Behold the happy home' in *The Maid of Honor*, and a more sectional style in 'Sweet spirit of my thought' in *The Sicilian Bride*. By contrast (judging from *Letty, the Basket Maker*, the later version), *The Devil's In It* fell back on clichés--the normal ballad-style, and polka and waltz rhythms--in the search for box-office appeal at the Surrey Theatre, where Bunn made his last stand as a manager.

THE FINAL PHASE

Satanella

Prior to his final period of composing for the London stage, Balfe's most recent opera had been given in Italy--*Pittore e Duca*, of 1854. He had now to adjust to the English conditions. He risked no more grandiose experiments with recitative, for instance. Apart from his operetta *The Sleeping Queen*, however, the six operas, plus the unfinished *The Knight of the Leopard*, mark an expansion compared to his earlier works. Their scale is larger, and the vocal capabilities of the Pyne and Harrison company gave him a freedom in writing for the voice that he had never consistently enjoyed in England. Balfe's first opera of this period, *The Rose of Castille*, of 1857, was a vehicle for Pyne's skills. Charles Santley, who made his Italian debut that year in *La Traviata* in Pavia, joined the company in 1859, and exerted considerable influence. Likewise, George Honey's talents generated certain buffo parts, such as Ralph in *The Puritan's Daughter*.

Balfe incorporated material from *L'Étoile de Séville*, of 1845, in the first two operas of this period. Zaida's chanson mauresque 'Je suis une fille maure' became the rondo mauresque 'O were I the Queen of Spain' in *The Rose of Castille*. Having been conceived for a 'chanteuse à roulades,' it was now suitably virtuosic for Pyne. The overture of the French opera became *Satanella*'s overture. A more general sign of

Parisian influence is the use of the galop in the slave-market scene in Act III of this opera.

Satanella's main musical theme, 'The Power of Love,' was sketched ten years earlier as part of Balfe's most intriguing uncompleted project, a setting of *Le Roi s'amuse*--another *Rigoletto*. His biographer, William Barrett, suggested that Balfe discontinued this work because 'his moral nature revolted at the equivocal nature of the subject'--another example of the partisan biographical style of the day (why did he start work on it?).[30] The sketch (Add.32672, ff.23v-24) has the melody as part of a duet in E major, in 4/4 (7.29.a). Balfe changed the key to D and gave the melody a more flowing character by rewriting it in 12/8 (7.29.b). The date of the sketch explains a similarity to 'When other lips' from *The Bohemian Girl*--the pre-cadential harmony and melody.

Satanella is representative of Balfe's late style in its richer use of orchestral effects and the renewed use of Italian elements such as the polonaise rhythm. The latter traits may be attributed to Balfe's recent period in Italy and to encouragement from Louisa Pyne. (In *The Rose of Castille*, the polonaise pattern appears under the guise of the Spanish bolero.) Pyne's skills are especially in evidence in Act III of *Satanella*. Her ability, and the enlarged scale of the opera in general, are made clear by recalling the equivalent scene for Arline in *The Bohemian Girl* ('Come with the gipsy bride,' in Act II). 'The Fascination Scene' consists of three movements for the soprano: a lively 6/8 allegro, with choral interludes; a 4/4 andante with obbligato viola, which concludes with a cadenza for voice and instrument, giving Pyne a top D;[31] and a fast waltz (3/8, allegro), again with choral interjections, and rapid passagework for the soloist.

An important element of *Satanella* is Balfe's broad grasp of the whole, as opposed to the effect of discrete units often found in earlier works. This will become clear if we consider the plot. It is a manichaeistic conflict played out by human and supernatural characters, focusing on opposing models of Womanhood and their influence on Man. The focal character is a Fallen Woman (Satanella), ordered by the Devil (Arimanes) to gain the soul of the pleasure-seeking hero, Count Rupert. Rupert is faced by his alluring but fickle fiancée Stella, who deserts him when he loses at the gambling table, and the loving Lelia, who is anxious to curb his profligacy.

Satanella falls in love with Rupert, but this love is portrayed as perverse, for she machinates the removal of Stella and of Lelia, whom

Rupert has now decided to marry, and takes Lelia's place at the marriage ceremony. However, she is discovered, due to beneficent supernatural intervention. With Satanella's assistance, Rupert rescues Lelia from a Tunisian slave-market; in return, he must soon surrender his soul to Arimanes. When the fateful day arrives, Lelia offers her own life to save Rupert. Overwhelmed by this demonstration of true love, Satanella decides to brave Arimanes's wrath and, sustained by Lelia's rosary, enters the protection of Heaven. In short, the theme of the drama is the redemptive power of love.

Balfe gave musical definition to this theme and to the major characters. The musical idea associated with the power of love recurs in each act, always linked to Satanella, and with a significant use of D major. As noted earlier, the composer sometimes changed his mind about keys, on occasion probably at the singer's request[32]--but in *Satanella* he seems to have had a schematic association in mind. The initial appearance of the theme, which occurs when Satanella is first made aware of the power of love, is in D. The succeeding appearances move away from this key when she draws closer to evil (to C and E in the introduction to her Act II aria, and to E flat in the marriage ceremony), and return to D in Act III as the redemptive power of love emerges. It is only in the D major appearances that she actually sings the melody (elsewhere it is in the orchestra), her words emphasizing 'the Power of Love.'

Minor keys tend to be associated with Arimanes's powers: his Act I aria (A minor), Satanella's reacceptance of her role as his slave in her Act II aria (C minor), and his invisible spirits in Act III (A minor). Also, the material is quite different. The arias for Arimanes and Satanella are faster, more forceful in their melodic angularity, and energized by the Italian cabaletta rhythm of quaver, two semiquavers, and quavers in the accompaniment. The music of the spirits' chorus is diametrically opposed to Satanella's theme in its three-bar units and bare scoring (7.30).

However, such features are not applied consistently. The melody with which Rupert opens the Act I quartet is similar to 'The Power of Love,' but at this point he is under the sway of an antithetical force; and his stirring call 'Up and arm ye, ev'ry brave' in the Act II finale is in the schematically appropriate key of D, yet utilizes the Italian rhythms otherwise reserved for Arimanes and Satanella. Two pragmatic factors were influential: Balfe preferred the Italian idiom for fast, energetic movements requiring a strong dramatic impact, and Harrison wanted a part (Rupert) that would neither strain his declining powers nor be conspicuously weaker than the other main roles.

Nonetheless, much of the characterization is of a superior quality compared to the earlier operas. Arimanes's aria 'Since he has dar'd to brave' confirms this (7.31). The vocal line of the introduction is more imaginative, the word-setting more satisfactory, and the orchestral figurations alter in tandem with the verbal phrases. In the aria itself not only the orchestral polonaise rhythm but also the vocal line catch attention. The rhythm and leaps of the latter's two-bar phrases are powerful gestures. The initial diatonic harmony soon becomes chromatic, a progression that is highlighted in the central episode by Satanella's response, which is of an entirely different character melodically, tonally (at first it is in D, the key of 'The Power of Love'), and as regards accompaniment. One senses a melodramatic villain brought full-bloodedly to life in music--a far cry from the portrait of Moncenigo in *The Daughter of St Mark*, created by the same singer, Weiss.

Elsewhere in the opera the harmonic style is more sophisticated, as is witnessed by the unexpected chromatic colouring in Satanella's aria movement 'Oh! could I but his heart enslave' and in her ballad 'Let not the world disdaining.' Rupert's 'No prize can fate on man bestow' is another sign of the loosening of the conventions of the ballad. Expanded to three verses, the second is set in a contrasting tempo and with a richer accompaniment (7.32).

The larger units are underpinned by a firmer sense of structure. The gambling scene in Act I and the Act II finale illustrate this: the sections follow each other with tighter effect. The Act III sextet and chorus, marked by a rousing unison melody, includes the only examples in the opera of the old rhythm of dotted-quaver-semiquaver, triplet--a surprising use of an outmoded pattern. The duet for Satanella and Arimanes reveals Balfe's strengths and weaknesses. In its first movement, Arimanes's response is strongly distinguished from Satanella's exposition. But the word-setting is very poor, and is worse in the second movement: '*his* proud, *his* proud' is obviously wrong, yet Balfe allowed it to pass. Such lapses indicate the variability of the composer's craftsmanship.

Other Operas of Balfe's Final Period

Compared to *Satanella*, the degree of Italian influence in the other operas of Balfe's final period is moderated. Among the examples are Wolf's 'cabaletta' 'Can it be' in *The Puritan's Daughter* and the Duchess's 'Passion's black and murky night' in *The Armourer of Nantes*, where Balfe used a wide-leaping vocal line and polonaise rhythms to illuminate her burning desire for revenge (7.33).

Making an exception for the movements involving Pyne, in some instances he seems to have taken even less care over characterization now that his career was re-established. Seymour's ''Tis hate, 'tis hate that fills my breast' in *The Puritan's Daughter* is perfunctory (7.34); and there is no difference in the musical expression of Blanche's rejoicing in Nature in *Blanche de Nevers* (marked 'la spagnuola') and Marie's denial of having a sexual liaison with the villain in *The Armourer of Nantes* (7.35). This is redolent of the relationship of music and text maintained by Rossini's generation, forty years previously. Yet Balfe was now faced by the new world of *Rigoletto*, in which there was a different approach. He was Janus-faced about such matters. The duet for Fabio Fabiani and the Jew in *The Armourer of Nantes* is better, according to the new criteria, as regards intention if not result. Its orchestral figurations change in step with the text, and the harmonic tension develops as the piece draws to an end. The unexpected use of speech set against music--one of the infrequent instances in Balfe's work--when Fabiani murders the Jew is imaginative. Despite this, however, the duet is not wholly satisfactory, for the majority of the orchestral patterns were chosen regardless of their appropriateness to the text. Similarly, the main movement of the Act I duet for Raoul and Fabiani cries out for different material for the two characters, but Raoul's bitter scorn--'The loathsome stain, the shiny trace'--looses its impact because Balfe gave the same music to Fabiani for his bemused incredulity at this outburst. The accompaniment for Marie's 'Oh! listen Raoul' in her duet with Raoul in Act III of *The Armourer of Nantes* has an effective countermelody for violas and cellos (7.36). However, the piece lacks that final degree of care: the effect is spoiled by the ungainly setting of 'ever,' in the last bar.

Balfe's better command of structure in larger units is exemplified by the concerted piece in Act I of *The Puritan's Daughter*. It is laid out in a tonal plan that highlights the conflict between the heroine and her father, expressed in duet movements: D flat (introduction), f-A flat (first duet movement), D flat (chorus), A (second duet movement), F (brief chorus), B flat-D flat (recitative), D flat (chorus). In the case of overtures, he altered his views: by 1861 the melodic style of *The Puritan's Daughter* was old-fashioned, whereas his last two operas, *The Armourer of Nantes* and *Blanche de Nevers*, have newer-style short preludes.

Two trends emerge as regards the ballad--the old 'Harrison' style, and the freer approach observed in *Satanella*. There is little basic difference between Harrison's 'Oh smile as thou wert wont to smile' in *The Daughter of St Mark* and ''Twas rank and fame that tempted thee' in *The Rose of Castille*, and the long A flat over diminished-seventh harmony in 'O love, thou'rt like a reed bent low' in *The Armourer of Nantes* is almost a repeat of the same point in 'When other lips' of twenty years earlier. The only change in Harrison's type was its increasing vocal restrictions--in 1861, his (Rochester's) ballad in Act III of *The Puritan's Daughter* was limited to the octave G-G. (The hero, Clifford, was played by Santley.) Balfe himself used the term less exclusively. In his manuscript score of *Satanella* he marked 'The Power of Love' 'ballad,' solely on the grounds that it is a slow-paced movement dealing with generalized emotions (and doubtless with an eye to sales: it was issued as a separate item). Pyne evidently wanted her ballads to have more dramatic impact and individuality, exemplified by Mary's 'How well I recollect the night' ('The Paradise of Love') in *The Puritan's Daughter* (7.37) and Marie's 'There's one who rear'd me' in *The Armourer of Nantes* (7.38). Mary's has a far more varied melodic shape than normal, with syncopations, contrasting short patterns, and more fioriture; the tempo fluctuates, and the accompaniment is varied to an unusual degree. The context of 'There's one who rear'd me, lov'd me' gives it a particular status. It is unique in English opera of this period because of the heroine's admission that she has 'fallen.' (Satanella is less explicit.) Musically, it is based on the traditionally simple mode of expression, but an expressive freedom of pulse distinguishes it from the normal ballad.

The unfinished *The Knight of the Leopard* was prepared for performance by Costa, who conducted it at her Majesty's in 1874 (it was entitled *Il Talismano*). William Barrett wished this to be seen as the 'noble outcome of [Balfe's] genius,' on which 'he dwelt with loving fondness, determined to make it the work of his life.'[33] Desmond Ryan extolled its virtues against the background of Wagnerism that was gaining ground in the mid 1870s. Wagner's style, Ryan maintained, 'possessed a charm for, and exerted an influence over, Balfe's latest efforts.... But the difference between the expositors of English and of German opera was immeasurable; one, full of natural resource, and teeming, Schubert-like, with inexhaustible melody; the other, striving beneath the cloak of clever contrivance to hide poverty of ideas and lack of spontaneous conception.'[34] Barrett and Ryan were grinding axes with a vengeance. The opera has some particular features--the polonaise rhythm is absent, for instance, and some of the inner orchestral parts are a little more chromatic--but the basic conception of the heroine's role (Edith) shows little advance on Catarina's in *The Daughter of St Mark*. The Act II finale is as Italianate as ever. The glee ('A song to merrie England') was a rare concession to the popularity of such ensembles. Even Ryan was forced to admit that the opera has flaws.

SUMMARY

Balfe's first English opera was performed within a few weeks of Bellini's death, and he himself died in the year of *Die Walküre*'s first performance. Within that period, his contribution to English nineteenth-century opera was immense. He was the dominant member of the group of composers who, in the 1830s to 1860s, developed English opera from its state earlier in the century, in which original music played a small role, to the much larger works common by the 1860s. His contemporary, Bayle Bernard, described him as 'almost the Atlas, it may be said, on whose shoulders our modern English opera reposes.'[35] More than any other of his colleagues, Balfe's success on the continent helped to spread the reputation of English composers abroad.

Balfe's early experience of the works of Rossini and his followers shaped his musical outlook. But it is noteworthy that the most Italianate of his operas are those that occur at the outset of each of the three phases of his career. External forces exerted a powerful influence, and while he habitually turned to the world of Italian opera for finales, for example, during the course of each phase Balfe moderated the Italian flavour. He constantly adjusted his style in accordance with the prevailing conditions. In pragmatic terms, this was essential in order to retain his position as the leading British operatic composer; but such compliance is a corollary of his weakness as a musical dramatist. Fitzball maintained that simple melodies would always rank higher for English audiences than 'the Trouvatores and Traviati,'[36] and, as Bernard Shaw pointed out, herein lay Balfe's strength--not in dramatic writing.[37] Meyerbeer's judgement of *The Bohemian Girl* was essentially the same: 'basically it is all melodic, [and] of course there are a lot of trivialities and plagiarisms under the surface.'[38] Lack of originality was a frequent criticism. Balfe's view of this matter may be judged from the following:

> He was disputing with a friend on the merits of a new German opera, which the latter, among other things, praised as eminently original. "Original?" exclaimed Balfe, "and what do you mean by that?" "Why, I mean to say that it is music which was never heard before." "Well, and I must say," he replied, "that it is music which will never be heard again."[39]

There speaks the pragmatist.

Dramatically, Balfe was inferior to some of his English contemporaries--Benedict and Loder especially. He did not make a sufficient effort to hone his technique into a sharp dramatic tool. He never seems to have been fired with the strong conviction that led a Verdi or a Wagner to fight for his beliefs. Balfe's yardstick was melodic and theatrical effectiveness, not dramatic truth. He lacked that willingness to take pains for the sake of artistic principle that marks the great operatic composer.

We may turn to the candid Macfarren for the final word, for his appraisal of Balfe went straight to the heart of the matter. He praised Balfe's role in raising the status of English composers at home and abroad and acknowledged his facility of invention, but, with characteristic candour, he also remarked that

> against these great advantages is balanced the want of conscientiousness, which makes him contented with the first idea that presents itself, regardless of dramatic truth, and considerate of momentary effect rather than artistic excellence; and this it is that, with all his well-merited success with the million, will for ever prevent his works from ranking among the classics of the art.[40]

NOTES

1. Regarding the title and details of the performance of this work, see Ifan Fletcher, 'Balfe and Barrymore at Milan,' *Theatre Notebook*, 7 (1953): 60-62. The biographical details given here are mostly drawn from George Macfarren, *The Imperial Dictionary of Universal Biography*, edited by J.F.Waller, 3 vols (London, Glasgow, & Edinburgh, n.d.), s.v. 'Balfe, Michael William,' I:357-58; Kenney; and William Barrett, *Balfe: his life and work* (London, 1882).

2. See Kenney, p.223.

3. See Bellini's letter of 17 September [1829] to Monteleone, in Luisa Cambi, 'Bellini. Un pacchetto di autografi,' in *Scritti in onore di Luigi Ronga* [edited by Riccardo Ricciardi], (Milan & Naples, 1973), p.64.

4. See Carlo Gatti, *Il Teatro alla Scala*, 2 vols (Milan, 1964), 2:37, 38. Kenney's implication that Balfe sang Dandini and Figaro, not only Iago, at La Scala (Kenney, p.99) is not confirmed by Gatti.

5. See *Songs, Duets, Trios, &c. as sung in the Concert, at the Royal Gardens, Vauxhall, on Friday, August 14, 1835, for the Benefit of Signor Venafra* [London, 1835].

6. The opening words also occur in the cavatina for Corrado in Verdi's *Il Corsaro*--an example of the way in which librettists (Piave, in Verdi's case) availed themselves of a common stock of expression.

7. For Harrison's repertoire, see Willert Beale, *The Light of Other Days*, 2:230-31.

8. My translation of the letter, dated 29 nov[embre] [18]44, now in the Pierpont Morgan Library, to which I am grateful for supplying a copy. Punctuation is as in the original. The 'Gipsy' refers to *The Bohemian Girl*. The reduction from four to three acts occurred at a late stage: gauging from Balfe's autograph MS score (see BL, Add.29343, ff.136-37), it appears that the final scene was originally considered as Act IV. In the letter Balfe initially referred to the opera as 'Catarina Cornaro,' the title of Donizetti's version.

9. See White, *A History of English Opera*, p.282.

10. Fitzball, 2:25.

11. Fitzball, 2:19.

12. William Makepeace Thackeray's sketch of 'Signor Balfi' in the role at Drury Lane is reproduced in White, *A History of English Opera*, p.267.

13. Three decades later Balfe contemplated using this solo to expand the role of the queen in *La Bohémienne*, the French version of *The Bohemian Girl*, but did not pursue the idea. See BL, Egerton 2741, ff.28-28v. It is not clear whether this sketch relates to the performances at Rouen (1862) or Paris (1868).

14. *Athenaeum*, no.501, 3 June 1837, p.404.

15. For a more laudatory opinion of *Falstaff*, see Hurd, 'Opera: 1835-1865,' *The Romantic Age*, p.320.

16. For an indication of her Italian style of singing, see Balfe's 'Una donna più felice.'

17. Quoted by Kenney, p.159.

18. See Balfe's letter of 29 November 1843 to G.A.Osborne, quoted in the latter's 'Musical Coincidences and Reminiscences,' *Proceedings of the Musical Association*, 9 (1882-83): 103.

19. Willert Beale recollected that the initial reaction on the first night was mixed (*The Light of Other Days*, 2:229). It is not clear whether he implies there was a hostile claque. See also Clifford Harrison, *Stray Records* 2 vols (London, 1893), 1:96, 104-6, where Harrison also reports Balfe's similar trouble and care over the drinking song for Rochester in Act II of *The Puritan's Daughter*. Harrison's assertion that Balfe often borrowed ideas from Beethoven's sonatas and symphonies (1:108) is one of several examples of the partisan style of biography then current.

20. See the *Catalogue of Manuscript Music in the British Museum*, edited by Augustus Hughes-Hughes, 3 vols (London, 1906-09, reprinted London, 1964-66), 2:425, 432.

21. Bunn's original version of the text (see BL, Add.42970, ff.438v-39) was copied in the printed libretto. For Balfe's manuscript version, see Add.29335, ff.183-92. The vocal score noted here is that published by Chappell [1843?]. For the text as it appears in Chappell's score, see chapter 2, p.41.

22. See Sims Reeves, *My Jubilee; or, Fifty Years of Artistic Life*, 2:78-80.

23. For clarity, I use Bunn's names for the characters when referring to the French version.

24. *Athenaeum*, no.892, 30 November 1844, p.1099.

25. *Athenaeum*, no.891, 23 November 1844, p.1074.

26. Fitzball (2:208) and Kenney (p.185) wrongly suggest that Miss Romer played Arline in *The Bohemian Girl*.

27. *The Times*, 28 November 1844, p.5.

28. *Athenaeum*, no.892, 30 November 1844, p.1099.

29. The recitatives that Balfe wrote for performances of *Fidelio* when he was conductor at Her Majesty's show a different approach, incorporating motifs from the other sections (e.g. Jaquino and Marzelline's Act I duet in their recitative immediately preceding the Act I finale). In the Italian version of *The Bohemian Girl* (*La Zingara*) he superimposed voice parts above the 'melodramatic music,' producing a parlante effect. *Falstaff* has some recitative using figured bass, probably for piano (see BL, Add.29334, ff.155, 207, 335). This was in line with the somewhat conservative practice at Her Majesty's: see the score used for performances of Federigo and Luigi Ricci's *Il Nuovo Figaro* in 1837 and 1838 (BL, Add.30914, 30915--for instance, Add.30914, f.79).

30. Barrett, p.201.

31. Willert Beale relayed the anecdote that during the rehearsals of *Satanella* Balfe proposed adding a cadenza to the soprano's melody 'The Power of Love.' 'Alfred Mellon, while conducting the band, remarked that he was sure Miss Pyne could sing any phrase of a florid character any instrumentalist could play. Hereupon Pratten, the flautist, and Lazarus, the clarionettist, improvised the most intricate passages they could contrive.... The prima donna immediately took up the challenge, and imitated them on the stage with an accuracy and precision which amazed and delighted Balfe and all the orchestra.' (*The Light of Other Days*, 2:235).

32. To other examples may be added the instance of the heroine's 'Scena ed aria' in Act IV of *Bianca, the Bravo's Bride*. This was originally written in the sequence D minor to D major, but Balfe instructed the copyist to transpose the second movement up to E flat. (See BL, Add.29357, f.245v.)

33. Barrett, p.252.

34. Kenney, p.269.

35. Bernard, *The Life of Samuel Lover, R.H.A.*, 1:124.

36. Fitzball, 2:208.

37. See *Shaw's Music*, 2:718.

38. Giacomo Meyerbeer, *Briefwechsel und Tagebücher*, edited by Heinz Becker and Gudrun Becker, 4 vols (Berlin, 1960-85), 4:185, entry for 28 January 1847 (my translation).

39. Bernard, 1:125.

40. Macfarren, *The Imperial Dictionary of Universal Biography*, 1:358. Macfarren's comments were quoted by Edward F.Rimbault in *A Dictionary of Music and Musicians (A.D.1450-1889)*, edited by Sir George Grove, 4 vols (London, 1899), s.v. 'Balfe, Michael William,' 1:128.

VIII

Wallace

Career and Output

Few musicians have had such a colourful career as Vincent Wallace (1812-1865). This 'excellent *Eccentric man*' enthralled Berlioz (a firm admirer of his most popular opera, *Maritana*) with tales rivalling the *Arabian Nights*.[1] Indeed, so flamboyant was his life that the major biographical problem is separating fact from fiction.

Wallace first became known as a violinist in Dublin. He was also an outstanding pianist. In 1835 he went to the Antipodes, moving to South America in 1838. During the next few years he made his way northwards. It seems that in 1844 he travelled through Germany and Holland. He was in London in 1845. *Maritana* was produced that autumn. Wallace later resumed his travels to North and South America, returning to Germany in 1858. He moved to London in 1859, and his last operas were produced here in the early 1860s. In 1864 he fell seriously ill, and died in France.

Six of his eleven operas were performed.[2] *Maritana* (15 November 1845) and *Matilda of Hungary* (22 February 1847) were both given under Bunn's management at Drury Lane. After a hiatus came *Lurline* (23 February 1860), *The Amber Witch* (28 February 1861), *Love's Triumph* (3 November 1862) and *The Desert Flower* (12 October 1863). *The Amber Witch* was produced at Her Majesty's, and the others by Pyne and Harrison's company at Covent Garden.

Wallace's operas are divided into three groups by chronology and style. We shall consider them in this light, paying most attention to *Maritana* and *Lurline*.

Maritana and *Matilda of Hungary*

Maritana is set in Madrid. The eponymous heroine and Don Caesar de Bazan are caught up in courtly intrigue inspired by the villainous Don Jose, who wishes to alienate the queen from the king of Spain. Thanks to the assistance of the warm-hearted youth Lazarillo, Don Caesar survives an attempt on his life, saves the monarch's honour, and gains Maritana's hand.

In his review Chorley remarked that *Maritana* incorporated 'half a dozen different manners tried in as many portions of the opera.'[3] The quip has an element of truth. One of the episodes in the overture, which alternates themes from the opera with passages of bustling activity, is a mock fugue on the well-known diminished-seventh subject. Opposed to this academic extreme is the unexpected Neapolitan harmony combined with a strange fourth above the bass in Maritana's romance 'It was a knight' (8.1), an exotic effect that Nicholas Temperley has suggested may have been noted during Wallace's travels.[4] There is abundant use of the bolero rhythm to inject colour into this 'Spanish' work, and a waltz--the latter being unexceptional by 1845.

Nevertheless, the basic style is Italian. Don Jose's aria 'This heart by woe o'ertaken' supplies a clear illustration. Comprising Rossini's pattern of a cantabile immediately followed by a cabaletta, its slow movement includes a near-literal quotation from 'Io sono docile' in Rosina's aria 'Una voce poco fa' in *Il Barbiere di Siviglia*: compare 'last hope forsaken,/ Sinks in despair' and 'prima di cedere farò giocar' (8.2). Italian elements abound in the ensembles, such as the Bellinian parallel lines and rhythms of the duet for Maritana and Lazarillo, 'Ah! Sainted Mother,' in which 'And thy gentlest pity lend him' may be compared with 'ah! perchè la vuoi scemar' from 'Mira, o Norma' in *Norma* (8.3). Equally Italianate is the design of the Act II finale, with its central 'ensemble of perplexity' ('What mystery'), in which the stunned soloists sing successive entries in the style of a Rossini canon. For all its quiet beginning, the 'Angelus,' a choral prayer, has the swinging rhythm and full-blooded sonorities of an Italian chorus (8.4). The orchestral figurations, too, are frequently Italianate in their rhythms and texture, as in the parlante 'Of fairy wand had I command' (8.5).

There are three ballads, two with texts by Bunn, 'In happy moments day by day' and 'Scenes that are brightest,' and one by Fitzball, 'There is a flower that bloometh.' The latter is in the mould of Balfe's 'When other lips' in *The Bohemian Girl*. Sung by the same singer, Harrison, it is in the identical key, D flat, and has the smooth lines of its predecessor. Wallace gave more prominence to 'Scenes that are brightest' (8.6), inserting it in the overture and at the beginning of the Act III duet for Maritana and Don Caesar. Marked 'Cantabile e con molta semplicità,' it reveals a fine sense of lyricism. The repetitive rhythm, always emphasizing the down-beat, could have been clumsy in weaker hands.

Matilda of Hungary is similar in style to *Maritana* but much less convincing. One suspects it was composed in haste. The Italian element emerges in the frequent use of the polonaise rhythm and in features such as the three-part form of the Act II finale--not least remarkable for one of Bunn's weakest lines, Podiebrad's statement that 'Unable though for her to live,/At least for her I'll die.' However, Wallace's word-setting was not unimpeachable: consider 'They who would *still* be happy *must*/ But little *in* the future *trust*' in Matilda's romance in Act II. The three ballads are weaker than *Maritana*'s. In his efforts to contrive simple-sounding, balanced phrases, the composer did not maintain sufficient rhythmic variety.

Lurline and *The Amber Witch*

These operas differ from the foregoing works in their less exclusive reliance on Italian idioms, their fuller use of the orchestra, more thoughtful use of recurrent material, and greater depth of characterization. They are of a more serious mood, fully in accord with the nature of the plots. Certainly the latter are of a sober tone. The earlier work, for instance, concerns the love between Lurline, the siren of the Rhine, and Rudolph, who is forsaken by his inconstant fiancée, Ghiva. Lurline's father, Rhineberg, is persuaded to allow Rudolph to return to his castle laden with riches. The avaricious Ghiva is now so anxious to conclude a match with Rudolph that she throws a ring the nymph had given him into the Rhine. Distraught, Lurline arrives to upbraid Rudolph, but is driven by love to save him from his false companions, who are drowned in the river.

In 1848 *The Musical World* confidently announced that *Lurline* was to be mounted at Covent Garden by a company Bunn hoped to form, with Sims Reeves and Eliza Wallace, the composer's sister, among the performers. The plan came to naught. Wallace's French biographer, Arthur Pougin, believed the opera was performed in Germany in 1854,[5] but all the evidence points to the performances at Covent Garden in 1860 as being the first. By this time the circumstances of English opera had undergone changes that influenced the character of the work, particularly the advent of the Pyne and Harrison company. We may be sure that Louisa Pyne expected virtuosic passages in heroine's parts written for her and that Charles Santley required imposing baritone roles. Equal account needed to be taken of William Harrison's restricted capabilities.

The stages of the composition of the work are confirmed by details in the composer's autograph full score.[6] The bulk of Acts I and II was written in 1847.[7] Wallace made some additions in 1848, but then put the work aside until, we may assume, 1859-60, when the possibility of performances became a reality.[8] Several of the additions made in 1848 and 1859-60 are short transitional sections such as recitatives, and not all were retained in the vocal score. The main alterations made in 1859-60 to Acts I and II concern revisions of the Act I finale and the insertion of a recitative and ballad, 'The nectar cup may yield delight,' for King Rhineberg--Santley's part--in Act II. Wallace wrote the latter in D flat. For some reason it was transposed to B in the vocal score, though flat keys were more common for ballads. A high C for the tenor in the Act I finale was left unaltered, but doubtless Harrison found an alternative.

Regarding Act III, Lurline's long scena 'Sad as my soul' probably dates from 1847, and Wallace began work on the final two numbers in September 1848[9]. However, most of this Act was written in 1859-60, including the ballad 'My home! my heart's first home' for Harrison, a recitative and andante ('Whence those looks with anguish fraught') for Santley, the orchestral passage representing the Rhine's deadly waves, and the virtuosic conclusion of the finale for Pyne. (Evidently she was content with 'Sad as my soul.') The overture--the only item not in the composer's hand--was written on paper of the 1859-60 period, but it is not possible to say when it was composed.

In cultivating a more serious musical style Wallace may well have been stimulated by other works that appeared in the intervening years. The publication of fantasias on *La Traviata* and *Rienzi* in 1857 and

1859 testifies that he kept abreast of operatic developments. Even so, we may not go so far as Pougin, who maintained that *Lurline* was influenced by Weber, Meyerbeer, Rossini, Donizetti, Bellini, Verdi, Herold, and Auber--thereby suffering, not surprisingly, from a 'lack of unity.'[10] Pougin was over-enthusiastic in his identification of styles, confusing specific influences with common stock, while paying too little regard to the conventional English features of the opera, ballads and glee-like ensembles. The latter comprise 'Though the world with transport bless me' and a hunting chorus. Likewise the waltz in the Gnome's Act II drinking song is yet another example of this dance's popularity. Pougin was correct, however, in his reference to Weber. The first subject of the overture's allegro is of a similar cut to that in *Oberon*, and there are further traces of his style in the use of 'melodramatic' delivery in the chorus and dance in Act II. Also, the concluding allegro of 'Sad as my soul' has more in common with Agathe's scena in *Der Freischütz* than with an Italian cabaletta (8.7).

Meyerbeer's influence is to be discerned, if at all, in connection with the more expansive orchestral effects. But here, too, Wallace's model is more likely to have been Weber. The overture's sober introduction includes far more ambitious use of the woodwind and chromatic part-writing than is found in *Maritana* (8.8). The orchestral portrayals of the river Rhine are especially interesting. The Introduction of Act I ('andante ondeggiando') has a horn melody and cello triplet arpeggios to illustrate its wave-motion, and upper woodwind for birdsong (8.9). The whipping-up of the Rhine's waters in the Act III finale is another instance of colourful instrumental writing (8.10). Again a precedent occurs in *Oberon*, introduced by similar words: 'Wind and waves! obey the spell!' in Weber's chorus (in Act I of *Oberon*), and 'Waves their destruction prove' in *Lurline*. Bearing in mind that it dates from 1859-60, the passage also raises the question of Wagner's impact--specifically, the seascapes in *Der fliegende Holländer*. While we cannot be categorical, there remains the possibility that this passage is the earliest instance of Wagner's influence in English opera, if not in British music as a whole.

The accompaniments of the vocal sections are generally richer than in Wallace's previous operas, adding to the expressive power of the music. For example, Lurline's description in her second romance, 'When the night-winds sweep the wave,' of the doom to which she draws unhappy mariners is marked by chromatic orchestral effects (8.11). Another instance occurs in the Act II finale. High celli, 'con dolore,' introduce Lurline as she begs her father to spare Rudolph's life--a moving plea that subtly changes from short, pleading motifs to more extended lines, doubled by oboe. Simultaneously an ominous bass trill figure associated with the enraged Rhineberg gives way to a higher-scored accompaniment, tinged with Neapolitan harmony (8.12).

The recall of material is more sophisticated. When Rudolph sings to his friends about the nymph of the Rhine, part of Lurline's romance 'Flow on, flow on, oh! silver Rhine' quietly appears in the orchestra (8.13). Lurline's prayer, first heard in the overture, reappears in the instrumental introduction to her scena well before the movement in which she sings the melody. A Wagnerian analysis would suggest that this indicates Lurline's mind was moving towards the idea of prayer before she actually voiced it.

Italian idioms are still present, but it is noteworthy that they are most prominent in the material written in 1847 and 1848. The fluency of melodies such as Lurline's 'Flow on, flow on' noted above and the shaping of the vocal and orchestral parts at the beginning of the duet for Rudolph and Ghiva, 'Lady, let this arm sustain thee' (8.14), are general examples. The Act II finale provides a more specific illustration. The section 'This lowly suit forebear' (8.15) is a concerted passage effectively in 9/8 rhythm, with a slow harmonic pace, in which the hero and heroine share a sweeping melody doubled by the upper orchestral parts. A comparison with 'È rimasto là impietrato' in the Act II finale of Donizetti's *Don Pasquale* of 1843 indicates the Italianate style. The passage is followed by the conventional stretta.

The quality of the ballads varies according to the importance of the roles. 'The nectar cup may yield delight' (8.16) has a broad sweep, well matched to Santley's vocal qualities, and Rudolph's 'My home! my heart's first home' supplied Harrison with his usual smooth, if vocally-restricted, solo (written in E major, its top note is G sharp). But Ghiva's 'Troubadour enchanting' is the weakest item in the opera. Added in 1848, it is one of a number of pieces probably included to bolster a secondary part, and Wallace devoted less care to it. Given the richer scoring of the opera as a whole, the ballads raise a particular problem: their slim accompaniments are incongruous.

The Amber Witch is set in Pomerania. The innocent Mary, known for the generous use she makes of a secret vein of amber, is assailed by the lascivious Commandant of the area. His evil servant Elsie incites the superstitious community to turn against Mary, who is found guilty of witchcraft. She is about to be burned at the stake when she is rescued by the hero, her beloved Count Rudiger.

The company at Her Majesty's included Helen Lemmens-Sherrington (one of Louisa Pyne's chief rivals) as Mary, Sims Reeves as Rudiger, and Santley as the Commandant. The conductor was Hallé. It was a strong cast.

Pougin did not include Wagner's name amongst those who influenced *Lurline*. In this case, however, he claimed to discern the influence of 'the dangerous theories of a very remarkable musician, but subject to wandering--I speak of M. Richard Wagner.'[11] His view of the latter would have been especially influenced by the Parisian performances of *Tannhäuser* in March 1861, a few weeks after the premiere of *The Amber Witch*. We should not assume that the Frenchman was correct. But certainly there is a stronger sense of dramatic coherence. A more serious tone develops as each of the four acts succeeds another. There is an appropriately bright quality in Act I. Act II consists of a massively-extended love duet whose sections interlock with a witches' chorus; the final choral section recalls the first, giving a sense of unity to the design and reinforcing the witches' dramatic role. Acts III and IV maintain the concern for continuity and sober mood, which Chorley felt was dissipated by the rondo finale that was imposed upon him, doubtless largely, if not entirely, by Lemmens-Sherrington. The *Athenaeum* declared that 'We are requested to state that the sprightly final *rondo*, which immediately succeeds to incidents so serious, was placed there in disregard of the protest of the arranger of the *libretto*.'[12] The limited use of Italian idioms is noteworthy. For instance, while 'Haughty maid, my passion scorning,' the last movement of the Commandant's scena in Act I, is marked 'Cabaletta,' the predictable polonaise rhythms are absent. Instead of an overture there is a bipartite Introduction, its first section based on a fanfare, its second consisting of a chorale; both recur at various points. The most effective recall of material concerns an orchestral bass motif (it is never sung) consisting of a sequence of intervals of a fifth. First heard at the appearance of Elsie and her vicious companions in Act II and occurring in each subsequent act, its relevance is clarified by Mary during her trial in the Act III finale, when she identifies the evil force she has encountered as 'Disdaimonia'--Greek for superstition (8.17).

The recitatives are scored more fully than in Wallace's early operas, continuing the trend seen in *Lurline*, and throughout the work there is greater concern for instrumental sonorities. The cello, alone or in tandem with violin or viola, is a favourite sonority (for instance, Rudiger's Act I recitative 'Had ever son' uses cello and violin), and the woodwind are also used to particular effect, such as the passage for three flutes and clarinet in Mary's recitative 'The cloud hath pass'd' in Act II. The plot did not call for the orchestral passages required in *Lurline*, but Wallace took the opportunity presented by the hero's incarceration at the beginning of Act IV to write a lengthy introduction, which, unexpectedly, is fugal.

There is a change, too, in the ballad. Mary's 'O never laugh, Sir Knight' (Act I), though titled 'Ballad,' is of a very different cast from those of Bunn's day. It is in 6/8, not the usual 4/4, and has a refrain for three voices at the end of each verse: a highly unusual procedure.

Love's Triumph and *The Desert Flower*

It is symptomatic of the conditions of the day that Wallace's final two operas are of quite a different style and scale. In *The Desert Flower* ballad-style songs rub shoulders with polonaise rhythms and virtuosic display for Pyne, plus a modal melody (8.18) and a rhythmic ostinato to characterize American Indians. *Love's Triumph* comprises twenty-three short numbers, compared to the forty-three substantial items in *The Amber Witch*. Planché derived the plot, which is set in France, from *Le Portrait Vivant*, a comedy by Mélesville and Laya of 1842. The text is witty and fast-moving, its light-heartedness epitomized by the musical phrase 'Carelessly cantering down a lonely dell' (8.19) as well as a classic statement of the contemporary view of Woman: 'Goddess! Woman! Angel! Devil!' in the air 'Wayward Fortune.' The librettist was angry at the way Pyne and Harrison cut the work to accommodate the annual pantomime.[13] Chorley was largely correct in his view that the work was French in idiom.[14] Wallace matched the mood with the light melodies, jaunty rhythms, short phrases, and neat, quick orchestral figurations of the opéra-comique style. But it is not exclusively Parisian. This is a catch-all work, catering to varied tastes. The polonaise rhythm is used ('Wayward Fortune,' marked 'Tempo di Polacca') and there are a minuet for period flavour (Act II finale) and waltz rhythms for contemporary appeal (the duo 'To the secret'). For Pyne there was the longest aria, 'O rank, thou hast thy shackles,' and for Harrison the ballad 'Lovely, longing, and beloved,' which does not rise above F.[15] 'Corin for Cleora' is a glee complete with 'Tral lal la' refrain.

SUMMARY

Wallace's operas are as varied as he was restless. *Maritana* is one of the most exuberant English operas of the nineteenth century. Its Italian flavour is unmistakable. In his last four operas, appearing one a year from 1860 to 1863, Wallace adopted other styles. *Lurline* and *The Amber Witch* are set on a different plane of seriousness. This partly reflects the mood of their libretti, but also involves a change in the composer's approach. His range of musical resources broadened, particularly in terms of orchestral colour. It is symptomatic that of all the British composers now studied only in Wallace's case does the possibility of Wagner's influence arise. Probably we shall never know how deeply Wallace was committed to this Germanic style or how much he would have developed it had he been presented with further opportunities, rather than having to turn to the lighter French-style works that were the last he was able to compose.

The variety of idioms represented in the composer's operas bears witness to his breadth of interests and versatility. The esteem with which Wallace was regarded abroad may be gauged from Berlioz's admiration and from the fact that another Frenchman, Pougin, wrote a biography within two years of his death. By contrast, it is to Chorley's discredit that he dismissed the final work of his erstwhile collaborator with the criticism that, like all his operas, it had too little local colour.[16] If there is one feature that marks Wallace's operas, it is their varied flavour, epitomized by the bolero in *Maritana*, the painting of the river Rhine in *Lurline*, the chorale in *The Amber Witch*, the Parisian sparkle of *Love's Triumph*, and the exotic melodic flavour in *The Desert Flower*. Only *Matilda of Hungary* falls beneath a certain standard. We may not agree with Pougin that English opera died with Wallace--'English opera is therefore dead, really dead'[17]--but, with his passing, the liveliest spirit of all the English opera composers of the period was no more.

NOTES

1. 'C'est un excellent *Excentric man.*' See Berlioz, *Les Soirées de l'Orchestre* (Paris, 1852), p.412. Berlioz also formed a warm friendship with Balfe, with whom he worked under Jullien's management during the season 1847-48. See Berlioz's letter to Joseph d'Ortigue, dated 15 mars [1848], in Hector Berlioz, *Oeuvres Littéraires Correspondance Générale*, edited by Pierre Citron, 5 vols (Paris, 1972-89), 3:528.

2. Those remaining unperformed were *The Maid of Zurich*, *The King's Page*, *Estrella* (unfinished), *Gulnare*, and *Olga* (the last two being Italian operettas).

3. *Athenaeum*, no.943, 22 November 1845, p.1130.

4. See Nicholas Temperley, *New Grove Dictionary of Music and Musicians*, edited by Stanley Sadie, 20 vols (London, 1980), s.v. 'Wallace, (William) Vincent,' 20:175-78.

5. *The Musical World*, 23 (1848): 482; Pougin, *William-Vincent Wallace: étude biographique et critique*, p.18. The possibility of performances at the Paris Opéra that same year has also been proposed (see note 4), but there is no firm evidence to support the suggestion.

6. The score was used by Alfred Mellon, who conducted the initial performances. I am grateful to Dr Paul Raspé, librarian of the Conservatoire Royal de Bruxelles, for his assistance during my visit to study the manuscript.

7. Wallace dated the first and last numbers of Act I 'Vienna 13 July 1847' and 'Vienna August 2 1847,' and the first number of Act II 'October 3d, 1847'.

8. The paper used at this period is watermarked 1858. Temperley states that Wallace assigned the copyright of *Lurline* to Pyne and Harrison on 18 March 1859 (see note 4).

9. Wallace dated these 20 and 23 September 1848 respectively. He was now at St Helier, Jersey.

10. Pougin, p.33.

11. Pougin, p.19.

12. *Athenaeum*, no.1740, 2 March 1861, p.299. Over a year later, in his review of *Love's Triumph*, Chorley continued to fulminate against such demands: 'There will never be a musical stage worth having in any country till artists cease to demand the same stereotyped occupations, the same *encores* in the same places,--till the tenor, be he

of what standing he may, will content himself without having any fatally-sweet ballad allotted to him,--till the heroine can be made happy without the fruitlessly-brilliant *rondo*,--till the bass will do his duty without his scene of stupid or sinister vengeance.' (*Athenaeum*, no.1828, 8 November 1862, pp.598-99.) The case emphasizes Miss Rainforth's likely role in the decision to alter the ending of *The Bohemian Girl*, which similarly ensured that she had a rondo finale (see chapter 3, p.53).

13. Planché, 2:222-23.

14. See note 8 above. Chorley tended to identify a wide range of music as French. His review of *Love's Triumph* continued (p.599): 'French effect is half the secret of even Signor Verdi's *cabalettas*, with their skips and their syncopations and their surprises ... the one single tune in Herr Wagner's Mediaeval 'Lohengrin,' where *Elsa's* maidens disrobe her on her bridal night, might have been signed "Adolphe Adam"'.

15. A top A near the end is ossia.

16. *Athenaeum*, no.1877, 17 October 1863, p.503.

17. 'L'opéra anglais est donc mort, et bien mort' (Pougin, p.38).

IX

Benedict

Career and Output

Born in Stuttgart, Sir Julius Benedict (1804-1885) received a thorough grounding in German music. At the age of fifteen he became a pupil of Hummel in Weimar, and met Beethoven while still in his teens. In 1821 he began his studies with Weber, an association that lasted until 1824, when he became conductor at the Kärntnertor theatre, Vienna. Benedict moved to Naples in 1825, and for nine years was one of the music staff at the San Carlo and Fondo theatres.

In 1834 Benedict travelled to Paris and the following year to London. He established himself as conductor at the Lyceum from 1836-37 and from 1838-48 at Drury Lane, under Bunn's various administrations. In 1848 he conducted Mendelssohn's *Elijah* with Jenny Lind taking part, and later toured America with her. He resumed his operatic conducting in 1852, at Her Majesty's. Benedict was also conductor at the Norwich festival for over thirty years (1845-78). He was knighted in 1871.

For most of his professional life Benedict was best known as a conductor and pianist. Opera composition occupied only a small part of his career. His earliest operas were three small-scale works performed in Naples: *Giacinta ed Ernesto* (about 1827), *I portoghesi in Goa* (1830), and *Un anno ed un giorno* (1836). He wrote five English operas, the first three being performed at Drury Lane: *The Gypsy's Warning* (19 April 1838), *The Brides of Venice* (22 April 1844), and *The Crusaders* (26 February 1846). After almost twenty years came *The Lily of Killarney* (8 February 1862) and the small operetta *The Bride of Song* (3 December 1864), both at Covent Garden.

Benedict's vocal works also include a number of cantatas, an oratorio, songs, and partsongs.

The Lily of Killarney is by far his most important contribution to English opera, and the present discussion is primarily devoted to this work. We shall approach it by considering aspects of his other operas.

THE EARLY OPERAS

Judging by *Un anno ed un giorno*, in Naples Benedict wholeheartedly adopted the idiom expected by Italian audiences. The work is replete with conventional turns of melodic and orchestral writing. His first English opera, *The Gypsy's Warning*, however, is markedly different. The Italian element is limited to a trio in the style of a Rossinian canon in Act II. By contrast, Beethoven's influence is clear. Benedict may have hoped to impress the critics at the opera's forthcoming production in Mainz in 1838. The overture opens with a remarkably chromatic opening passage (9.1), after which comes an allegro in C minor, redolent of the 'Pathetique' sonata in its first subject, the 'Eroica' in its fugato development, and the Fifth Symphony in its final C major blaze of glory. An additional point of interest is the recapitulation's reversal of the order of the first and second subjects. It is possible that Benedict had encountered Berlioz's *Symphonie Fantastique*, with its formally similar first movement, during his stay in Paris.

The ballads betray an inexperienced hand. 'Scenes of my youth' and 'He comes not yet' are melodically constricted and jerky in rhythm, with double-dotted crotchets and semiquavers, as opposed to the popular flowing style. The most expansive solo is the soliloquy for the villain (Ludovico), 'Rage thou angry storm.' Its main melodic material also appears as the second subject in the overture. A dramatic orchestral introduction (again in C minor) and recitative are followed by a strophic moderato. The latter includes a change from B flat minor to major to express Ludovico's delight in his evil passions, but it is a disappointing piece, due to the small scale and the lack of vitality of the ideas.

The overture of *The Brides of Venice*[1] has not survived. The glee-like ensembles and ballads are predictable for the period. Benedict had still not caught the lyrical flow required for the latter. Nor are the arias satisfactory, for they lose dramatic momentum. Three consist of a recitative followed by a single movement, as in *The Gypsy's Warning*, while the 'cavatina' 'O memory cease to grieve me' has the Italian form of a cantabile and a cabaletta. The latter has a stilted effect. The ballad-writing in *The Crusaders* is slightly more fluent, but it is impossible to discuss this work in detail as only six items survive.

The Lily of Killarney

By 1862 Benedict's ability as an operatic composer had matured greatly. *The Lily of Killarney* is one of the most accomplished of all English operas. Quite apart from its unusual Irish setting, the intrinsic quality of the musical ideas, the powerful characterization, and the control of dramatic momentum give the work a special place in the history of British theatrical music.

It will be recalled that the plot concerns the landowner Hardress, who is secretly married to the peasant girl Eily. In order to retrieve his family fortunes he is pressed by his mother Mrs Cregan to marry the wealthy Anne Chute. His servant Danny Mann, in his own terms ever faithful despite having been crippled by Hardress, hopes to make Hardress's advantageous marriage possible by murdering Eily. The heroine is saved by the adoring Myles, and Hardress acknowledges 'the Colleen Bawn' as his wife. The parts of Eily and Myles were created by Louisa Pyne and William Harrison and Danny by Charles Santley.

Two particular features characterize the musical style of the opera. One is the influence of Weber. Although he had avoided following the older composer's style in his earlier operas, Benedict's admiration for his former teacher remained strong. (Eventually he was to write the important study *Weber*, published in London in 1881.) In 1860 he renewed acquaintance with *Der Freischütz* when writing recitatives for a revised version performed at Her Majesty's. In *The Lily of Killarney* 'melodramatic' delivery of the text is used frequently, most prominently in the Act II 'Melodrame,' where what would otherwise have been the one duet for Danny Mann and his intended victim is carried out in speech accompanied by music (9.2). This is also the most concentrated point for the recall of themes associated with elements of the drama. The thematic recurrence is more advanced than in any other English opera of the time. The associative material includes themes relating to Myles (a particular song sung by the character) and to the happy denouement, but special attention is paid to crucial elements of the plot--the signal that Hardress will visit Eily, her eager anticipation, the danger she faces, and Danny's murderous frame of mind. The latter two are solely orchestral. Three of the themes appear in the overture, whose opening moderato alternates an ominous dotted figure with the motif denoting the threat to Eily. There follow an allegretto based on Myles's song and Eily's anticipation, and then an allegro which focuses on the happy outcome. The choice and sequence of motifs is not as synoptic of the ensuing drama as in the overture to *Der Freischütz*, which Benedict admired greatly,[2] but Weber's model is obvious.

The other special musical element in *The Lily of Killarney* is folk-like material. Benedict used this to distinguish the Irish peasantry from the local gentry. The quartet 'Let the farmer praise his grounds' sung by Eily, Sheelah, Myles, and the priest Father Tom is a notable example (9.3). Based on the Irish air 'The Cruiskeen Lawn,' a setting of which (by W.M. Herbert) had been published in 1857, it consists of two statements from Myles alternating with a refrain. Myles's unsophisticated solo sections are accompanied by basses that have a Baroque air--affording a fascinating glimpse of Benedict's vision of a musically undeveloped world--while the emphasis on the flattened leading note in the refrains lends a modal quality. The instrumental introduction to Myles's first recitative also has a modal tinge, and the vocal line has a simplicity and lyrical quality that are quite different from the recitative of an Italian opera, for example (9.4). Myles is potently characterized by his recurrent song 'It is a charming girl I love,' whose vocal line and matching accompaniment have the quality of a country tune (9.5). Making little concession to Pyne's virtuosic abilities, Benedict personified Eily by similar means, but her music has a thicker texture and subtlety commensurate with the position of heroine. In her romance 'In my wild mountain valley' (9.6) the delicate modal colouring of the first strain of the melody contrasts with the more tonal second phrase, with its old-style 4-3 cadential suspensions. The repetitive pattern of quaver and two semiquavers in the voice and orchestra further distinguish the song. Such characterization is enhanced by Eily's most affective solo, 'I'm alone' (9.7). The avoidance of the leading note, the ambiguous cadences in bars 4 and 8, and the flattened leading-note harmony for the climax of the song all add effect.

A consequence of Benedict's concern for appropriate musical expression is that Eily's cry 'he comes, he comes, he comes o'er the waters to me' (in 'In my wild mountain valley') did not result in a large aria modelled on Agathe's similar enthusiasm at the approach of her beloved in *Der Freischütz*. Almost all the solos are single movements. But within this limitation Benedict created an imaginative range of musical characterizations. The contrast between Eily and Anne Chute is striking. Both of Anne's solos--in the hunting chorus at the beginning of Act II, and her air 'The eye of love is keen' (9.8)--are far more florid, indeed coquettish, than Eily's turns of phrase, with wide leaps and rhythms and accompaniments that are more sophisticated. Likewise the boisterous choruses of Hardress's friends that introduce Acts I and II are in a tonal vein far removed from that of the Irish quartet noted earlier. The Act I introduction derives from Italian opera in its form and rhythms. The contrast of idioms is made clear when Hardress interrupts Myles's quartet: orchestral patterns and harmonies revert to the norm from the pseudo-archaic. Hardress's appearances are usually in ensembles. His one independent solo is his ballad 'Eily, mavourneen.'[3] This is among the best of all ballads of the period (9.9). Benedict achieved the required simplicity in exemplary fashion. The rhythmic repetition in the vocal line supplies the desired effect, while the harmonic variety provides a counterbalance.

The crux of the drama is Danny Mann's belief that he can best serve his beloved master by murdering Eily. He reaches this decision in his duet in Act II with Mrs Cregan and in his ensuing aria. In composing Danny's part Benedict departed from his pattern of characterizing the peasantry by folk-like material. Acknowledging Charles Santley's status, he wrote the duet and aria in an Italian idiom. The duet opens with rhythms common in Rossini's scores and a parlante section whose orchestral motif is very similar to that in the duet for Lucia and Enrico ('Appressati, Lucia') in Donizetti's *Lucia di Lammermoor* (9.10). There follow 'Oh, Colleen Bawn, your reign is over,' a long and tonally intricate solo for Danny, a brief transition that includes the theme associated with Eily's danger, and finally a stretta for both characters. The latter ('No blot on our scutcheon shall e'er have a place') is of martial mood, with melodic lines in the shape of trumpet-calls and with familiar orchestral rhythms (9.11). Danny's aria is even more conventionally Italian in form. After the usual preparatory movement with its mixture of recitative, arioso, and orchestral figuration come a lyrical cantabile, an interlude, and a cabaletta. The andante ('The Colleen Bawn from childhood I have known') is the jewel of the opera (9.12). The harmonization is noteworthy. The first phrase's progression--tonic to supertonic--was no novelty: it occurs in 'Casta Diva,' in Bellini's *Norma* (1831). In order to conclude sequentially on the tonic, however, the answering phrase begins on the chord of the flattened leading note. This is not modal writing but the result of a wide range of harmonic colour, enhanced by chromatic inner part-writing. But the remainder of the aria lacks this dramatic cutting-edge. The emotive lines 'Down, coward scruples, hold thy peace, remorse!' were allocated to the tempo di mezzo, while the last movement, 'Duty, yes, I'll do my duty' (9.13), for all its dotted rhythms and dynamics, dissipates tension instead of heightening it. Despite its exploitation of Santley's wide vocal range, the curiously Handelian passage 'Danny Mann' (9.14) indicates that Benedict's old weakness, finding adequate expression for a villain at fast tempo, had not been overcome.

The attempt on Eily's life takes place during the Act II finale. It begins with 'melodramatic' delivery (9.15). Eily's spoken question 'What place is this you have brought me to, Danny?' is followed by the theme intimating her danger. Danny recounts how he was crippled by Hardress, his vocal line gradually rising while he relates his tale. As the confrontation between the two characters develops the tonality is intensified from D minor to an ambiguous G sharp minor. Danny attempts to drown Eily, but a shot rings out: and he falls into the water (9.16). The tension is relaxed, and to the strains of an A major melody Myles enters and soon rescues Eily. The passage building up to Danny's attempt on the heroine's life makes too regular use of repeated units, but the entire episode reveals an imaginative mind. Altogether, there is much to admire in this opera.

It would be unjust to judge Benedict by his last operatic work, *The Bride of Song*. Requiring only four soloists (there is no chorus), this operetta was too restricted in size to allow him to emulate the standard of *The Lily of Killarney*.

SUMMARY

Benedict's talents as an opera composer emerged only in his maturity. Then, spurred by a good libretto, he produced a work whose quality is rare in the English repertory. The focus on the main dramatic issues, clearly illustrated in the characterization of the four main personalities--Myles, Eily, Danny, and Hardress--and the emotional impact of the crucial scenes involving Danny testify to Benedict's grasp of dramatic effect.[4] We should not overlook his achievement in matching the ambience of the text to musical idioms. His use of folklike material went well beyond a mere revival of the late eighteenth and early nineteenth-century practice. The characteristic music is used with genuine dramatic point. *The Lily of Killarney* represents a considerable contribution to nineteenth-century opera as a whole.

NOTES

1. Benedict provided Bunn with some material for the text of this opera. See Bunn's preface to the libretto, p.iv.

2. See Julius Benedict, *Weber* (London, 1881), pp.68-69.

3. Commercial songs focusing on an Irish heroine were well established. F.N.Crouch's *Kathleen Mavourneen*, for example, dates from about 1840. (Balfe's *Killarney* was published in 1864).

4. Three variants may be noted. A manuscript orchestral score in the library of the Royal Academy of Music includes an alternative finale (3:ff.212-20) in E major (the original rondo finale is in A), using Hardress's ballad 'Eily mavoureen.' It consists of sections for chorus, homophonic quartet, and chorus again. Boosey's edition of 1872 ('Royal Edition' series) contains an appendix containing a substitute for the Act II duet for Myles and Eily, and an 'Irish National Dance,' 'introduced in the marriage scene, and which is now printed, through the courtesy of Sir J. Benedict, for the first time.' One of the melodies of the dance (it is sectional) is marked 'St. Patrick's Day.'

X

Macfarren

Career and Output

The career of Sir George Macfarren (1813-1887) was unique. Of all the prominent composers of English opera in the period, only he was a student at the Royal Academy of Music. Eventually he rose to exalted posts in the establishment, succeeding Sterndale Bennett as Principal of the Royal Academy of Music in 1875 and becoming Professor at Cambridge in the same year. Macfarren was knighted in 1883. By this time he had long been blind, having lost his eyesight during the composition of the opera *Robin Hood* (1860). His articles on musical subjects rank him as the most explicitly thoughtful British composer of the day, the musical Ruskin of his time.

Macfarren's work includes incidental pieces for various musical plays, commencing with *The Prince of Modena* of 1833. His operatic abilities are seen to better advantage in *The Devil's Opera* (English Opera House, 13 August 1838), *Don Quixote* (Drury Lane, 3 February 1846), and *King Charles II* (Princess's, 27 October 1849). After a long gap came *Robin Hood* (Her Majesty's, 11 October 1860), *She Stoops to Conquer* (Covent Garden, 11 February 1864), and *Helvellyn* (Covent Garden, 3 November 1864). He also produced two chamber operas--'opere di camera'--performed at the Gallery of Illustration: *Jessy Lea* (2 November 1863) and *The Soldier's Legacy* (10 July 1864). *El Malechor* (1839), *Allan of Aberfeldy* (1851), and *Kenilworth* (of about 1880) were never performed. The fate of *El Malechor* is an extreme instance of the result of the theatrical instability of the period. It was accepted by Bunn at Drury Lane in 1839, Barnett at the St James's in

1840, and Balfe at the Lyceum in the same year, but in each case the downfall of the management prevented it from seeing light of day. *Allan of Aberfeldy* was accepted for performance at Drury Lane in 1851, but met a similar fate.

Macfarren's father, George Macfarren senior (1788-1843), exercised an important influence. He was an experienced man of the theatre and author of a number of stage works, including *The Devil's Opera* and *An Adventure of Don Quixote*. He became proprietor and editor of the *Musical World* in 1841. In the composer's words, George senior regarded 'the operas of Mozart, Beethoven, and Rossini as the standard of plan and limit of extent for the construction of a lyrical drama.' His suggestions 'were a priceless schooling in dramatic composition ... his prompting of what to think, write, and avoid, made me an apprenticeship of which I, and none but I, can feel the inestimable value.'[1] He did not live to see *An Adventure of Don Quixote*, for, though started in 1839, seven years elapsed before it could be performed. The opera was dedicated to his memory.

Also highly influential were Macfarren's years at the Royal Academy of Music, which he attended from 1829 to 1836. He studied with Cipriani Potter, who had been a pupil of Attwood--through whom Macfarren could thus trace his lineage to Mozart. Another mentor, Woelfl, had been in contact with Beethoven. Of the younger generation of German composers Macfarren had most admiration for Mendelssohn. The relationship was reciprocal. Mendelssohn championed Macfarren's works in Germany, choosing his overture to Planché's musical drama *Chevy Chace* and a symphony for a concert in Leipzig in 1843, while it was Macfarren who conducted his friend's incidental music for Sophocles's *Antigone* at Covent Garden in 1845.

Macfarren's style developed over a period of twenty-five years. In order to follow this process we shall consider aspects of his early style as seen in *The Devil's Opera* and *Don Quixote*, and then investigate his mature work.

MACFARREN'S EARLY PHASE

The Devil's Opera and *Don Quixote*

These two works illustrate the variety of Macfarren's early compositional style. The first reveals the strong impressions left from his student days, while the second shows his response to influences stemming from the circumstances of performance.

Seen alongside contemporary operas of Barnett and Balfe, the striking feature of *The Devil's Opera* is its unusual choice of stylistic models. The influence of Mozart and Beethoven emerges strongly. This is particularly so of the overture. Its clear sonata form is Mozartian, while the ideas themselves reveal Beethoven's impact. Seven years later, spurred by identical admiration, Schumann used a similar idea for the finale of his Second Symphony (10.1).

The solo vocal items--small-scale single movements--also have a Viennese quality. The melody of 'Queen of Cities,' for instance, bears a likeness to the slow movement of Beethoven's violin sonata op.30 no.2 (10.2). Even the ballad 'O blame me not' has a classical cadence, with multiple suspensions. Traces of other foreign influences are few. Contemporaries would have regarded the canonic ensemble 'Good night!' as a Rossinian feature; doubtless Macfarren would have preferred comparison with 'Mir ist so wunderbar' in Beethoven's *Fidelio*. (Certainly Banister linked it with Beethoven's quartet.)[2]

Macfarren made a careful choice of folk material to colour the Venetian setting, hunting out the monfrina, 'a National Dance of the Venetian Peasantry.' This concern for appropriate music was typical. In 1836, when writing the *Chevy Chace* overture, he commissioned his brother to search for the English tune of that name.[3]

The plot of *Don Quixote* doubtless commended itself to the Macfarrens because Mendelssohn had set it in 1827 (as *Die Hochzeit des Camacho*). It was Macfarren's only opera to be produced under Bunn's administration at Drury Lane. It had previously been accepted by various managements that went bankrupt. The final version reflects the circumstances facing the composer. For instance, one part (Basilius), which had originally been conceived for Balfe, a baritone, was rewritten for a tenor, Allen, complete with high D.

It is a larger work than *The Devil's Opera*, and has a different stylistic character. The classical Viennese idiom has been left behind, and Macfarren made some concessions to contemporary Italian and German elements. For example, the tenor's solo in the finale of Act II, 'Hear me, great Heav'n,' has Italianate turns of phrase, and leads to a Rossinian canonic ensemble, 'All hearts relenting.' Some text is delivered 'melodramatically.' This feature of German romantic opera was by now unexceptional, as was the choice of the bolero and seguidilla to establish the Spanish setting.

Macfarren regarded characterization as important. He focused attention on Don Quixote with ridiculous roulades, trills, and falsetto effects, as in the finale of Act I (10.3). The virtuosic passages in Basilius's solo 'Life is an April Day' were probably concessions to the performer, and more characteristic of the work are Quiteria's heart-felt song 'Ah, why do we love' (10.4) and the ballads, such as 'Calm those frowning looks' (10.5). The former was part of the first version of the opera and had already become popular. Macfarren distinguished between song and ballad; the song has a more active vocal line and accompaniment, more colourful harmony, and a more complicated structure.

THE DEVELOPMENT OF MACFARREN'S MATURE STYLE

As his career progressed Macfarren cultivated two features that distinguish his operas from those of his contemporaries. These are a particular harmonic style and a concentration on English national elements.

Harmonic Style

Macfarren's harmonic practice was unique among English opera composers of the day. He adopted a conservative style that partly reflected the influence of national music but was also due to his theoretical views on harmony. He was a staunch advocate of Alfred Day's *A Treatise of Harmony*, published in 1845.[4] Essentially derived from Rameau, this is preoccupied with matters such as triadic formations and the building of discords above a root. Compared to contemporary continental developments the resulting harmonic idiom was outmoded, but the more abstruse points of Day's theories were controversial amongst English academics. In 1847 Macfarren resigned his teaching post at the Royal Academy of Music rather than foreswear support for the *Treatise* and only returned, in 1851, when he had an assurance of complete freedom in such matters. The composer's *Six Lectures on Harmony* of 1867 give a clear indication of his harmonic preferences. His models were still Mozart and Beethoven, and it was in their works that he saw the extremes of harmonic effect, as his discussion of the use of the submediant on a tonic pedal, a 'dangerous point on which to touch,' makes plain. He quoted two examples from the overture to *Don Giovanni* and the *Eroica* Symphony, adding:

> I must content me then with the bare statement of the fact that the two greatest masters of modern musical art have written what I have here described ... I leave the justification of such daring passages to other analysts, or at least till another opportunity.[5]

In view of this outlook it is not surprising that Macfarren's harmonic vocabulary is restricted. Though fully aware of 'the exhaustless field of the beautiful that lies open to the modern musician in the inexhaustible resources of chromatic harmony,'[6] he used such effects sparingly; the norm is diatonic, with a preference for common triads. On occasion the harmonic rhythm is unusually rapid. The vertical concept of harmony influences the texture and phrasing of the music, for it frequently moves in progressions of block chords and in regular phrase lengths.

National Style

We may approach this subject by considering Macfarren's attitude towards the use of foreign idioms. We are fortunate to have his own writings as a source of information. Around him he saw attempts to imitate Weber's style, but such attempts were in vain because it was inherently 'one that cannot be imitated (since all who have attempted its adoption have fallen into the most vapid musical bathos).' Rossini's idiom received strong criticism. Weber's 'best efforts were sorely hindered, if they could not be frustrated, by the ever poisonous working of the Italian predilections of his time.' Closely related was the matter of originality, an important concept for Macfarren. 'It is the unlikeness of the style of an author to any archetype that constitutes his originality,' he wrote. The 'most truly original composer the world has known' was an Englishman, Purcell (Macfarren edited *Dido and Aeneas*). But being original was not the sole yardstick by which a composer should be judged. Weber's music, 'however teeming with originality ... is greatly wanting in what may purely be termed classicality.' By contrast, Macfarren still revered Mendelssohn.

As we shall see, Macfarren did occasionally include features derived from Weber's works. Nevertheless he strove to establish a national operatic style marked by English plots and musical idiom. *Don Quixote* is the last of his 'foreign' works. All his remaining operas, including the 'opere di camera,' take place in England. He had acquired a knowledge of English folk music through his work on Chappell's collections of this material. He incorporated musical elements such as modal writing and country dances to imbue his works with an appropriate character. In historical terms this represents a return to the tendencies of the eighteenth and early nineteenth centuries, when operatic works with a strong British flavour were popular. But the scale and artistic conscientiousness of Macfarren's operas elevates them to a different plane. Given the foreign cultural hegemony over English opera, his preoccupation may be compared to that of the mid European nationalists--some of whom he pre-dated.

THE LATE OPERAS

Robin Hood

Macfarren pursued his enthusiasm for national music and subjects in the *May Day* cantata of 1857, scoring the final chorus for pipe and tabor. Its text refers to Robin Hood. Three years later came the opera of that name, the composer's first new theatrical work for eleven years. It established various features of his mature operatic style. He focused attention on the central dramatic issue by means of a recurrent theme, here Maid Marian's melody 'True love.' For the three extended solos (one each for the hero, heroine, and her father) the Germanic scena is preferred to the Italian aria. Elsewhere the soprano's part includes figuration whose scales and arpeggios have more in common with eighteenth-century vocalizations than the contemporary virtuoso passages.

As in his earlier work, Macfarren gave much thought to characterization. The part of the sompnour, a malicious ecclesiastic, is marked by the accompaniment of suspensions redolent of church music. The contrasting and forcefully English idiom that marks Robin and his associates is a major feature of the opera. Robin's 'Englishmen by birth are free' is uncompromisingly different from an Italian or German model in its simple style, as is the slow movement, 'Pow'r benign! the wish fulfill,' of Marian's scena ('Hail happy morn'), with its pentatonic contour (10.6). More national colour emerges in the long finale of Act II. Beginning with the chorus 'How bright is the day,' set to a dancelike melody (10.7), the movement proceeds to a Round Dance cast in the mould of a baroque chaconne--a clear indication of Macfarren's leaning towards the past (10.8). It is followed by episodes that, again, are essentially English: 'Tilting at the Quintain' and 'Hoodman Blind' (a version of 'Blindman's Bluff').

In 1863 Macfarren's 'opera di camera' *Jessy Lea*, in effect an operetta, was produced at the Royal Gallery of Illustration managed by Mr and Mrs German Reed. The following year saw the last of his operas: *She Stoops to Conquer* and *Helvellyn*, at Covent Garden, with another 'opera di camera,' *The Soldier's Legacy*, at the Marylebone Theatre, intervening. We shall concentrate on the latter three, turning first to the operetta.

The Soldier's Legacy

This is a short work. Set in Staffordshire in 1814, it concerns the arrival of the hero, who had promised a fellow soldier that he would safeguard the dying man's child. Only later does it transpire that this child is the heroine. Meanwhile the two have fallen in love, and despite various complications they are eventually assured of a happy future.

The piece reveals a skilful command of the comic vein, with a quick dramatic pace to match the ethos of the operetta. Again a recurrent motif emphasizes the crux of the drama, the dead man's legacy. Its sequence of predominantly common chords, largely in root position and adorned only by a cadential suspension, characterizes the rural simplicity of those involved (10.9). But the outstanding feature of the work is its harking back to the style of *The Beggar's Opera* by introducing old melodies. Indeed, of the eight such melodies, each of which is identified in the vocal score, two--'When the heart of a man is oppressed with cares' and 'Cold and raw'--are variants of tunes also found in Gay's ballad opera. Macfarren chose the material on the basis of straightforward verbal references, exemplified by the appearance of 'Cold and raw' when Jack recalls bad weather. The use of the melodies differs from that in *The Beggar's Opera* insofar as they are heard only when a particular character, Christopher (the heroine's amorous guardian), is present, and they sometimes appear independently of the vocal line. This is illustrated by the case of 'Here we go round the Mulberry bush' (set in the minor, not major, mode) in the trio for Jack and Lotty (the hero and heroine) and Christopher in Act II (10.10).[7]

She Stoops to Conquer

Presumably Macfarren felt such specific use of national material was best suited to the smaller proportions of this 'opera di camera,' for he did not attempt it in the two larger works of 1864. As regards *She Stoops to Conquer*, we should note that no other native composer would have contemplated setting an English literary classic. Set 'one hundred years ago,' to quote from the vocal score, the operatic version of Goldsmith's play concerns the relationships between Marlow and Hastings and Kate and Constance, with Squire Hardcastle and the lumpish Tony Lumpkin in the background. The *Musical Times* remarked that the opera was 'thoroughly English,' in contrast with 'the mixture of styles to which we have latterly been accustomed at this national establishment [Covent Garden].'[8] The musical character of the opera is established by the overture. It begins with a modal opening phrase and continues in unwontedly colourful vein (10.11). The bare octaves and dancelike lilt catch the mood of the work, as do the drone basses in the Introduction of Act I (10.12). The opera may not incorporate actual folk material, but various melodies have a rural character. Tony Lumpkin's fox-hunting song 'In Grandfather's time in days gone by' (10.13) is an example: it is marked by simple patterns of pitch and rhythm and an unsophisticated accompaniment, with the occasional flattened leading note to give a slight modal twist. Later in the song come hunting calls to add flavour. There can be few, if any, other operas that include a cricket match (10.14) and a game of 'kiss in the ring.' The latter provided an opportunity for an English country dance (10.15). Macfarren caught the work's comic element well, as is illustrated by Tony Lumpkins's vacuous mimicking of Constance as she pretends to read a letter (10.16).

Regarding the harmonic style, the overture surpasses the rest of the opera in chromatic colour. The pedal point in the trio of the Act I Introduction, 'Ah, my girls!' (10.17) could have been taken from a page of Shield's theatrical works of the previous century. Indeed, the 4-3 suspension at the Squire's cadence in the same example is appropriate to the epoch of the plot. Nevertheless the generally restricted harmonic language impresses not as period colouring but Macfarren's preferred style. So chaste is the writing that the ballads, with their conventional simplicity, seem thoroughly and successfully integrated into the whole work, not set apart as in other English operas of the 1860s. Frequently the idiom is so diatonic that accidentals are rare occurrences, as witness the glee 'The Cuckoo sings in the poplar tree' (10.18) and Squire Hardcastle's song 'Once again my father's hall' (10.19). The melodic style of the ballads is also restrained. Thus the phrases of 'Why sadly sighs the ev'ning gale' consist of repeated short units (10.20) as opposed to the broader sweep cultivated by Macfarren's colleagues. The use of a fast harmonic rhythm to promote a jovial atmosphere ('Drink friends drink') is another aspect of his harmonic style (10.21).

The absence of foreign orchestral figurations is striking. There are, however, some points where Macfarren did include standard operatic idioms of the day. There is a small amount of 'melodramatic' delivery of text, most prominently at the beginning of Act III (10.22). Accommodating the soloists called for more care. Due to the nature of the plot none of the solos needed to be extensive; all are single movements. Even so, Louisa Pyne required a degree of exuberant passagework while Harrison's ailing voice was to be husbanded,[9] a contrast evident in the duet she sang with the tenor, 'I say madam, I say' (10.23). However, instead of the type of virtuosic writing encountered in an Italian or German opera, some of the soprano's semiquaver passages are in broken thirds, again recalling English music from a bygone age.

Helvellyn

The Times noted that the plot of *Helvellyn* was derived from a German source--Mosenthal's *Der Sonnenwendhof.*[10] Nevertheless, it is typical of Macfarren to have chosen a plot set in Wordsworthian country in the aftermath of the Industrial Revolution. The plot focuses on Hannah, the heroine, who has returned to the area in which, years ago her father had been held responsible for burning an iron foundry. He had died in the blaze. Hannah feels herself cursed by her unhappy background and does not allow her identity to become known. The other main characters are Mabel, Martin, the object of Mabel's admiration, and Luke, Mabel's villainous brother-in-law. Hannah becomes the focus of animosity from all but Martin. Forced to live in a wind-swept hut, she receives unwelcome attention from Luke. Suddenly childhood memories come to the fore: she recognizes him as the true arsonist. He had been motivated by revenge against the foundry owner, having been rejected as a prospective son-in-law. A storm arises; to protect himself from the gale Luke clings to a tree, but the tree is struck by a thunderbolt--and the villain falls to his doom in the abyss below. No longer burdened by a sense of fate, Hannah feels free to reveal her love for Martin. The magnanimous Mabel relinquishes all claim to his affection so that the hero and heroine may live happily thereafter.

Chorley reported that *Helvellyn* was poorly attended.[11] He was far more impressed by Gounod's *Faust*, now being performed by Pyne and Harrison's rival company at Her Majesty's. The critic ridiculed *Helvellyn*'s plot and described Macfarren's music as

> not the best specimen of his manner, made up of many manners--now, it must be presumed, past any rational hope of change. There is a little of everything in it ... Meyerbeer ... [Weber's] 'Euryanthe' ... Mr. W. Chappell's collection [of country dances] ... Signor Verdi ... Arne ... Halévy ... Auber: in the more ambitious pieces Mr. Macfarren seems to have been often at a loss; and to have written merely to get over the ground.

This is a typical exaggeration from the *Athenaeum*'s reviewer, though with its grains of truth.

Helvellyn is a larger work than *She Stoops to Conquer* (there are four, not three, acts). Its overture is unique in English opera of the period. It was planned to have a special role in the drama, adumbrating the plot by visual and musical means. Banister described it as follows:

> The introductory prelude was termed an "Illustrated Overture," the chorus singing behind the scenes, during its continuance, a maledictory theme, recurring in the opera, "Curse on the head that the evil planned." An important feature was to have been the lime-lighted *tableau* shown in the course of this overture, illustrating the progress of the story. This, for reasons of economy and convenience, was, however, abandoned for a drop-scene, which had in its centre a scenic representation of the murder by fire referred to in the malediction.[12]

This integration of musical, visual, and dramatic stimuli is the closest English opera of the period came to Wagner's concept of 'Gesamtkunstwerk.' It is salutary to note that the blind Macfarren could not have expected to see it.

National music is prominent. The overture begins with repeated small units common in a folk idiom, soon giving way to modal material (10.24). The pastorale section of the Act I finale and the country dance in the 'Harvest Home Scene' in Act II maintain the ambience (10.25). The offstage church hymn in the Act III finale, with its simple vocal scoring and doxology-like melody, adds to the effect (10.26). Also, Macfarren paid unusual regard to correct declamation of the language, as in Mabel's 'We'll speak of this hereafter' in the Act I finale (10.27).

The harmonic style is not as restricted as in *She Stoops to Conquer*--appropriate for the more involved plot. Also, by this time Macfarren was not above taking hints from Weber. There is considerable use of recurrent material and of extended solos. After appearing in the overture, a modal theme associated with the original crime recurs at several points in the opera, for instance in the early narrative ballad 'The sky with crimson is overspread' and when Hannah reveals her true identity to Martin in the duet 'The gen'rous Mabel.' In the latter it appears in the bass of the orchestra, acting as an ominous background to the dialogue (10.28).

The more expansive revelations of personality take the form of aria and scena. Regardless of the former term, Italian figurations are absent. The scena is distinguished by its more fluid form and variety of orchestral figurations. In Martin's scena 'She loves me' differences between the freer and the more lyrical sections are minimized by the use of a flowing arioso style in the former passages. Likewise Luke's scena 'For nearly a month' is a mosaic of sections in which only the final passage approaches an extended lyrical quality. Hannah's aria 'On my heart a weight' illustrates Macfarren's preference for old-style virtuosic figuration in the soprano's part (10.29). It is this that Chorley must have had in mind when referring to Arne. Yet the outpouring of emotion in Mabel's spinning-wheel song in Act IV has nothing archaic about it. The repeated interruption of the song by a sweeping vocal line over a tremolo accompaniment is an effective procedure (10.30). This is the most dramatic expression given to an individual in the opera.

The simplicity of the ballads matches the inherent straightforwardness of the country folk. They have an extremely limited harmonic style--Martin's 'A blessing on the harvest fall' barely has an accidental. Some other movements are more chromatic, such as Hannah's prayer 'When my ample store I see.' Strangely, however, the confrontation between heroine and villain in the Act IV finale is harmonically tame: she cries 'Demon!' to the sound of a dominant seventh in C major.

The latter example points to the weakness of the opera as a whole. Macfarren used a wide range of compositional techniques, but the musical ideas do not have adequate dramatic power. The villain's scena is another case. Admittedly the composer cannot have been stimulated by the bland text. Nevertheless, for all his evident desire to respond to the various textual implications, none of the sections of the scena grips attention and there is little sense of culmination.

SUMMARY

Even from the early stages of his career Macfarren adopted an approach that distinguished him from his colleagues. Instead of following the accepted procedure--adopting modern Italian and German operatic styles--he drew upon Viennese classicism. His later work is predictably well-crafted and conscientious. Its weakness lies in the lack of a strong dramatic quality. It is for this reason that *She Stoops to Conquer*, involving a less serious engagement with fundamental emotions, is his most successful work.

We may remember with a jolt that *Helvellyn* was performed only a year before *Tristan*. It does Macfarren more justice to remember that

She Stoops to Conquer and *Helvellyn* preceded Smetana's *The Bartered Bride* (1866), one of the major nationalist operas of the century. Macfarren set himself the goal of establishing an operatic style whose plots and music had a strong national quality. In this sense he provides a link between the earlier English school that had culminated in Bishop and the following generations. Of all the six composers now studied, Macfarren had most influence on his British successors. The brisk style of his opere di camera may be considered a point of departure for Sullivan's operettas. (*Cox and Box* appeared only four years after *Jessy Lea.*) In more serious English music, too, one senses Macfarren's legacy. His careful word-setting was an example to the fastidious Parry and Stanford. His musical language is not thoroughly imbued with national features in the manner of the later, fully-fledged nationalists, but his pride in English material prepared the way for Vaughan Williams and Holst. Above all, his influence is to be felt in the restrained harmonic style that characterizes English composers of the late nineteenth and early twentieth centuries. The typical triadic melody and chaste harmonic vocabulary of 'Once again my father's hall' in *She Stoops to Conquer*, for instance, anticipates a case such as Variation VIII, 'W.N.,' in Elgar's *Enigma Variations*. It is fitting that Macfarren left his most indelible mark in the realm of harmony, a subject so close to his heart.

NOTES

1. Banister, *George Alexander Macfarren: His Life, Works, and Influence*, pp.44, 6.
2. Banister, p.53.
3. Banister, p.39.
4. Macfarren edited the second edition: see *Alfred Day's Treatise on Harmony, Edited, with an Appendix, by G.A.Macfarren* (London, 1885).

5. George Macfarren, *Six Lectures on Harmony* (London, 1867), pp.206-7.

6. This and the following quotations are taken from Macfarren's articles 'Mendelssohn,' *Musical World*, 24 (1849): 34-35; 'The Italian language: its evil influence upon music,' *Musical Times*, 14 (1869): 8. See also Banister, pp.158, 211-13.

7. On the sources and uses of the melodies, see also Nicholas Temperley, 'Musical Nationalism in English Romantic Opera,' *The Lost Chord: Essays on Victorian Music*, pp.152-55.

8. *Musical Times*, 11 (1864): 239.

9. See Banister, pp.204-5.

10. *The Times*, 4 November 1864, p.7.

11. *Athenaeum*, no.1933, 12 November 1864, p.642.

12. Banister, p.205.

Conclusion

We are now in a position to consider English opera of the years 1834 to 1864 from a wider perspective.

A brief recollection of the state of English theatrical music in the earlier part of the nineteenth century allows us to assess the contribution of Balfe and his colleagues to the English repertory. Especially in the early stages of their careers they were faced by entrenched antipathy towards the role of music in drama. They were always at the mercy of managerial insolvency, highlighted by the years in the 1850s when no theatre was regularly available in London. Yet, despite this, these composers brought about a radical change. No other thirty-year period in the history of English music saw such a sustained effort to institute national opera as a solidly musical form. To turn from Bishop's last pieces to those dating from the end of Balfe's, Wallace's, Benedict's, and Macfarren's careers is a salutary experience. The genre still accommodated spoken dialogue, but its musical element had developed almost beyond recognition. The majority of Bishop's theatrical works consist of incidental songs and glees, having minimal bearing on the plot. During the years 1829 to 1840, when his career as an operatic composer effectively ended, his most substantial contributions were little more than arrangements of continental favourites. From Barnett's *The Mountain Sylph* onwards English opera expanded in length and substance, with music playing an essential role in the dramatic effect. Extended arias and concerted movements became standard. The composers who used national musical material did so with far more sophistication than their predecessors. Moreover, in Balfe the British had a musician of international fame, known throughout Europe for *The Bohemian Girl*. Wallace, too, was highly regarded on the continent. We are not so overwhelmed with riches today as to be able to scorn such acclaim.

A feature that emerges very clearly from our enquiry is the extent to which opera reflected important elements of the contemporary society--its values, portrayal of woman, and understanding of human nature through the passions. From the musical point of view, the fundamental issue is the way in which composers coped with the lack of a strong indigenous operatic tradition. Only the ballad and the glee consistently maintained a national idiom, and English opera owes much of its unique character to these components. Beyond this, however, one is made forcefully aware of the impact of Italian music. The influence of Rossini, Bellini, and Donizetti was overwhelming, approached only by that of Weber.

Our study also focuses attention on the low esteem in which British music is traditionally held. Dismissive generalizations to this effect have a long history, involving as they do a national acceptance of cultural inferiority. The following statement about Victorian attitudes, taken from a recent study of romantic music, is an example:

> the new middle class, powerful in economic and political terms, was characterized more by a philistine puritanism than by cultural open-mindedness.[1]

The careful definition that such terms demand is beyond our brief, as is the deeper sociological investigation that is required. Such a study should distinguish between encouragement for the separate categories of sacred music, instrumental composition--the symphonic repertory, for instance--and opera. For it is clear that the English public avidly supported many new operatic works, written in a range of musical styles as wide as could be found in any European capital. Indeed they were highly style-conscious, not least due to Chorley's columns in the *Athenaeum*. As an antidote to the negative view so often encountered, we should pause to consider the following extract from an article that appeared in *La France Musicale* in November 1843. The author had attended Drury Lane and the Princess's. He confirmed that the English had acquired a low reputation among foreigners, but continued in a vein strikingly different from what might be expected. The report is worth quoting at length.

> British barbarism, as regards music, has virtually become a proverb; well! there is no accusation less deserved.... In both theatres the entry on the stage of each artist was saluted by triple salvoes of applause. You can imagine, after such a beginning, the tremendous degree [paroxisme] to which enthusiasm was raised during the performance. The Neapolitan public, as ready to burst into flame as the crater of Vesuvius at the moment of an eruption, is a

> public of statues compared to that in London, where claquers and what are known in France as free entries are unknown.... Women sit in the pit alongside the men, and they are no less demonstrative of their displeasure than of their enthusiasm. Their little cries often mix with the strong voices of the men and dominate them on account of their piercing and prolonged sounds. In London, when an artist is successful, at least he can boast of having support from *everyone*; in France and elsewhere one only knows the judgement of the *men*.
>
> At Drury Lane ... the prima donna, Mme Rainforth, is a distinguished singer, well-trained in Italy, who would not be out of place on the best operatic stages of Paris. M. Jules Benedict directs the vocal music and conducts the orchestra.... This knowledgeable musician carries out his duties with superior intelligence and admirable zeal. I left London on the eve of the performance at Drury Lane of a grand opera by Balfe, whose subject is taken from the *Gipsy* [*The Bohemian Girl*]; I attended a rehearsal of this work, which will certainly be an outstanding triumph for the composer.... The *Peri* has had an epoch-making success at Drury Lane. The English have feted Carlotta Grisi, as they feted Elssler, Taglioni, Cerrito; the vogue for Th. Gautier and Burgmuller's ballet has only increased by the twenty-fifth performance. It is also necessary to say that the scenery is of a richness that is entirely British ['d'une richesse toute britannique']. The stage-decorator's art is in its infancy at Paris, if one compares it with what exists in England, even in the smallest theatres.[2]

Even making allowances for the obvious exaggerations (Miss Rainforth certainly would have been out of place at the Opéra, for instance[3]), is this the world of a restrained, culturally-limited public? The picture presented by the French correspondent is of an demonstrative audience that supported opera and ballet with great enthusiasm. Clearly we must be wary of disparaging assumptions about English musical and theatrical life.

But the refreshingly different view put forward in *La France Musicale* did not gain ground. The later history of opinion of English opera continued to be marked by what the commentator Charles Salaman described in 1879 as 'the inane prejudice of English men and women against the native music of their own country,'[4] and by the time of Balfe's death two other factors were involved: the increasing hegemony of the Vienna-Leipzig axis and the new German school of Liszt and Wagner. In 1875 Salaman had expostulated about 'the musicians of Germany [who] ... by their writings and their performances ... are striving to force into undue prominence the compositions of the modern German, or the so-called 'higher development,' school of music.' The latter attracted growing numbers of British admirers. They included, for example, Clifford Harrison, son of the tenor for whom Balfe had written many a melody: it was Wagner's music that he eulogized.[5]

More than any other element it was the advance of German tenets, especially Wagnerism, that undermined the status of Balfe and his colleagues. The cry was for dramatic effectiveness and cohesion as opposed solely to melodic charm. As we have seen, Macfarren found fault with Balfe on similar grounds, but now the criticism took on a new dimension. In 1882 the ardently Wagnerite *The Monthly Musical Record* celebrated the first London performances of *Die Meistersinger von Nürnberg*, the *Ring* cycle, and *Tristan und Isolde* (all of which occurred that year) with lengthy articles on Wagner's work, complete with musical examples of leitmotives. By contrast, while acknowledging Balfe's popularity as a melodist, the periodical complained that in his operas 'the dramatic sense is so frequently violated [that] we are constantly shocked at the variance between the dramatic situation and its treatment.' Structurally, the operas resembled 'old-fashioned bed-quilts ... patchwork.' In sum, 'one feels obliged to offer an apology for him.' Equally telling is a comment in the periodical's report of the unveiling of the Balfe memorial in Westminster Abbey in October that year: 'the composers of the present hour are seeking after higher, purer, and nobler things; and a day will certainly come when the best of Balfe's music will cease to be a joy.'[6] W. Barclay Squire summarized the views of the contending forces in a perceptive article of 1885--a year after the London branch of the Wagner Society was founded.

> Balfe's life ... saw the palm of musical pre-eminence transferred from Italy and France to Germany. When the 'Siege of Rochelle' was written, Wagner was unknown. Forty years later, when 'Il Talismano' was produced, the only living Italian composer of eminence had proclaimed to a great extent his adherence to the principles preached by the German school. Thus it is that opinions differ so widely as to the merits of Balfe's music. To musicians who judge

> him from the point of view of the old ideal, his brilliancy, melody, and fertility of invention will entitle him to a place beside Bellini, Rossini, and Auber, while, on the other hand, by those who look for deeper thought and more intellectual aims in music, he will be regarded as a mere melodist, the ephemeral caterer to a generation who judged rather by manner of expression than by the value of what was expressed.[7]

In the succeeding hundred years Balfe has increasingly been considered a 'mere melodist,' to the point where his works are now never performed on the major London stages. Beecham's performances of *The Bohemian Girl* at Covent Garden in 1951 remain an isolated challenge to the prevailing establishment.

The obstacles to a serious revival of the operas of this period are several. One is the conventions of the libretti. We have a distrust of stereotypes and a preference for psychologically intricate characters. In this regard, it is necessary to develop our understanding of the principles on which the plots were written and of the characteristic vocabulary, with its references to the passions. Turning to the musical aspect of the repertory, the strength of the main composer--Balfe--was, as Barclay Squire observed, his melodic writing. If *The Bohemian Girl* or *Satanella* are to regain a place in public esteem it will probably be on this account. His most inspired melodies may be compared with many that are now frequently heard. But the abiding weakness of his works, the lack of a dramatic quality, will always hinder their revival. Great musical dramatists have been able both to transfigure the emotional intensity of a particular moment and to develop a cumulative effect over broad spans of time, ultimately encompassing an entire work. Balfe was unable to do this. The case of the ballad, that cornerstone of his original success, is symptomatic. Dramatic impetus can too easily be interrupted while the standard two verses in slow tempo are sung. The crucial factor is not the nature of the ballad itself--it is no more inherently undramatic than any other type of slow movement--but the sensitivity with which it is related to the wider context.

This point leads to a broader issue concerning English opera. Balfe and others frequently drew upon a variety of foreign idioms, sometimes within a single work. Often this results in a sense of stylistic incongruity--pointing to a failing in the composers, not to a generic weakness of the medium. Purcell, for instance, was so able to stamp disparate national elements with his own personality that they are welded together into a satisfying whole. In the nineteenth-century repertory, Balfe's *The Daughter of St Mark* fails in this regard where Benedict's *The Lily of Killarney* succeeds, despite including German, Italian, English, and Irish elements. The ideas are more substantial in themselves; they are carefully chosen to meet the situations of the plot; and they are absorbed successfully within the entire framework.

Indeed, of all the English operas of the period it is *The Lily of Killarney* that most consistently maintains a high level of inspiration and craftsmanship. The development of the plot and the characterization of individuals, primarily Danny Mann, allow this work to stand comparison with others now far more before the public eye. Given the usual assessment of English opera, it is interesting to discover how the London correspondent of the *Gazetta Musicale di Milano* described *The Lily of Killarney* when it was performed by Carl Rosa's company in 1876.

> This opera is in fact a real jewel, whether in terms of musical beauty or dramatic interest ... it is with astonishment that one wonders why it is not part of the Italian repertory, for which it has all the elements of a sure success. [It is] a work certainly not destined to be forgotten.[8]

The author was wrong in his prediction, but we should ponder over his evaluation. There is no justification for overlooking works of this stature. English nineteenth-century opera now requires further scholarly inquiry, brought to notice in a developing body of literature, in order that its potential interest and value may be realized. May it receive the attention it deserves.

NOTES

1. Arnold Whittall, *Romantic Music* (London, 1987), p.153.
2. *La France Musicale*, Sixième Année (1843): 376-77.
3. Willert Beale reported that 'the vocal powers of Miss Rainforth were limited, but the quality of her voice was pure and telling' (*The Light of Other Days*, 2:234-35).
4. Charles Salaman, 'On Music as a Profession in England,' *Proceedings of the Musical Association* (hereafter *PMA*), 6 (1879-80): 113.
5. Salaman, 'On Musical Criticism,' *PMA*, 2 (1875-76): 9; Harrison, *Stray Records*, 2:9-15.
6. *The Monthly Musical Record*, 12 (1882): 67, 241. For articles on Wagner's works see, for example, pp.[73]-79. The first of Wagner's operas to be performed in London was *Der fliegende Holländer* (1870), followed by *Lohengrin* (1875), *Tannhäuser* (1876), and *Rienzi* (1879). *Rienzi* was given in English, the others in Italian.
7. W. Barclay Squire, *Dictionary of National Biography*, edited by Leslie Stephen, 68 vols (London, 1885-1900), s.v. 'Balfe, Michael William,' 3:47.
8. *Gazetta Musicale di Milano*, Anno xxxi (1876): 353-54.

Musical Examples by Chapter

Ex.2.1

or by the power of imagination. Let us consider the passions in the following manner: first, the source of each; secondly, its natural branches; and, thirdly, its deviations, by which means we shall be able to distinguish at once the genuine emotions of human nature from those that have been the consequence of its depravity. For the sake of precision I will treat of them alphabetically and accordng to the fol lowing sketch, or

ANALYSIS OF THE PASSIONS.

SOURCES.	VARIETIES.	DEVIATIONS.
AMBITION.	*Desire of Power.* *Desire of Fame.*	Avarice. Envy.
ANGER.	*Indignation.* *Resentment.*	Retaliation. Revenge. Rage. Fury. Fretfulness. Moroseness. Surlyness. Hastiness. Sullenness.
ANTIPATHY, OR AVERSION.	*Natural Repugnance.*	Hatred. Malevolence. Rancour. Spite. Misanthropy.

CURIOSITY.

cont'd

PART III.] THE PASSIONS. 245

SOURCES.	VARIETIES.	DEVIATIONS.
CURIOSITY.	*Deſire of Information.*	Futile Curioſity. Diſhonourable Curioſity.
FEAR.	*Timidity.* *Terror.* *Horror.* *Awe.*	Cowardice.
HOPE.	*Rational Hope.*	Chimerical Hope.
JOY.	*Chearfulneſs.* *Mirth.*	Exultation. Falſe ſpirits, fictitiouſly procured. Malignant Joy.
LOVE.	*Self Love.* *Sexual Love.* *Storgé.* *Eſteem.* *Friendſhip.* *Patriotiſm.* *Philanthropy.* *Benevolence.* *Charity.* *Gratitude.* *Piety.*	Pride. Arrogance. Haughtineſs. Vain-glory. Vanity. Jealouſy.
SHAME.	*Baſhfulneſs.* *Diffidence.*	Shame of doing right.

R 3 SORROW,

cont'd

SORROW, or GRIEF.	*Melancholy.* *Contrition.* *Remorſe.*	Deſpair.
SYMPATHY.	*Pity.* *Terror.*	Vicious Sympathy.
WONDER.	*Admiration.* *Aſtoniſhment.*	

CHAPTER I.

AMBITION.

Sect. I. *Its Varieties and Deviations.*

THIS paſſion is the deſire of great things; or rather of thoſe things which tranſcend our preſent ſtate or attainments. It is an eſſential quality in man to aſpire; it marks the ſuperiority of his ſpirit above the reſt of animals; and, in no ſlight degree, indicates his immortality. Aſpire, my children! but let your ambition be fixed on thoſe objects that legalize the paſſion. Whatever tends to the exaltation of your nature is the legal object of Ambition. Caſt your eyes to the ſummits of intellect, and virtue; and ſtrain every faculty to accomplish the aſcent.

Ex.4.1

266
RHYTHM.
344.
INANIMATE.
Moderato.
INDOLENT, SLUGGISH, LANGUID.
Moderato.
A LITTLE MORE ANIMATED.
Allo. Moderato.
EXCITEMENT.
Allo. con Spirito.
PASSION, TURBULENCE, IMPETUOSITY.
ff
Allo. Agitato.
VEXATION AND ANGER.
Agitato.
FEAR, TREMBLING.
pp
AGITATION.
rinf.
rinf.
rinf.
Maestoso.
RESOLUTION, DETERMINATION.
Staccato.
sf.
sf.
Moderato.
MEEKNESS, GENTLENESS, INTREATING.
p

cont'd

Ex.4.1 cont'd

RHYTHM—PERIODS.
267
AMIABLE, INSINUATING.
Moderato.
p
GRACEFUL.
Andantino.
dol.
SUAVITY.
Andantino.
dol.
PRIDE, HAUGHTINESS.
Pomposo.
rinf.
rinf.
CHEERFUL.
Allegretto.
pia
CONTENTMENT.
Allegretto.
pia dol.
INFANTINE PLAYFULNESS.
Scherzando.
mez. Voce

Ex.5.1

Ex.5.2

Ex.5.3

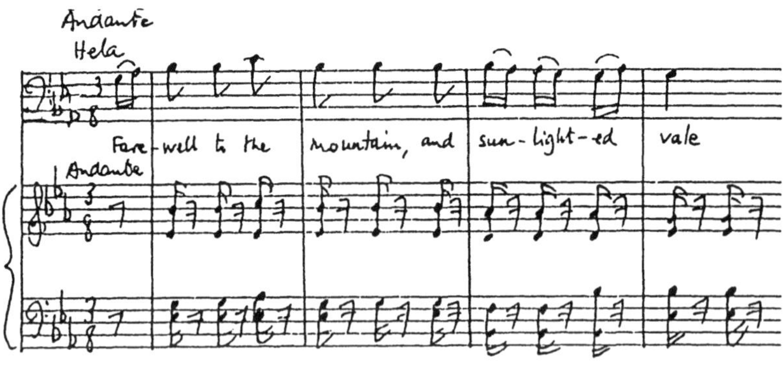

Ex.5.4

Ex.5.5.a

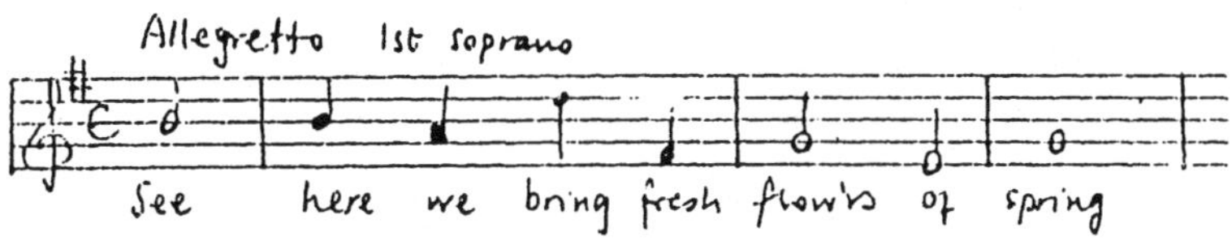

Ex.5.5.b

Ex.5.6.a

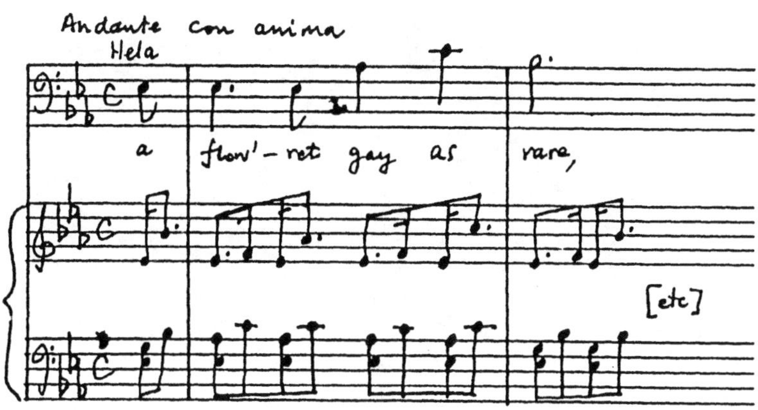

Ex.5.6.b

Allegro brillante
Hda
guilt re- joi-ces in des- pair
guilt re- joi— ces

Ex.5.6.c

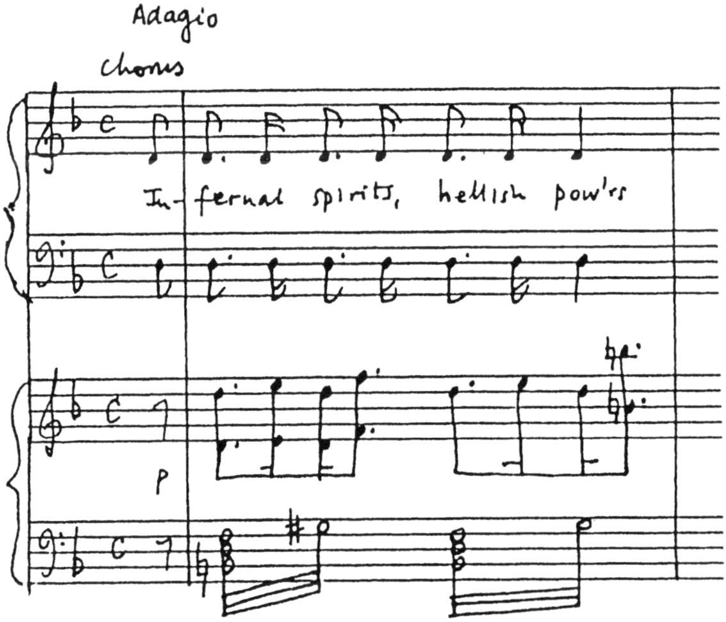
Adagio
Chorus
In- fernal spirits, hellish pow'rs
p

Ex.5.7

Ex.5.8

cont'd

Ex.5.8 cont'd

Ex.5.9

cont'd

Ex.5.9 cont'd

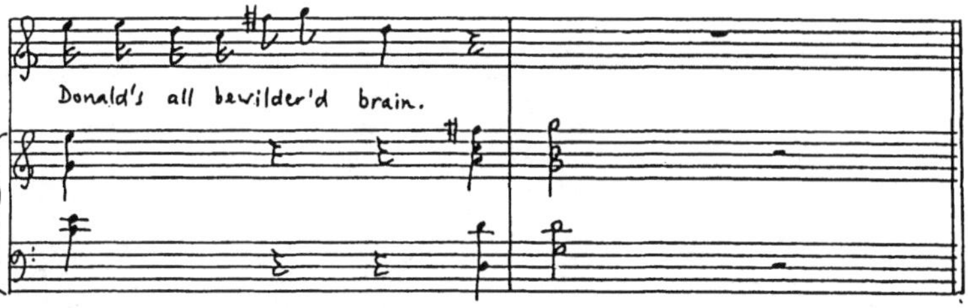

Ex.5.10

cont'd

Ex.5.10 cont'd

Ex.5.11

Ex.5.12

Andantino con moto
Jessie: My Donald's vows are sin- cere, My Don-ald's
Donald: My vows to Jes- sie are sin- cere, my vows to
Hela: trem - ble trem- ble, trem - ble fear, trem - ble,
f
ff
p
cresc.
lento
Allegro con spirito
Jessie: vows are all sin- cere.
Donald: Jes- sie are sin- cere.
Hela: trem- ble, trem- - - - - - ble fear.
ff
ff

Ex.5.13

Andante con anima
Recit.
Henry: The morning breaks! the ruddy east has wide its portals
flung, and forth the glor-ious sun in smiling radiance comes to wake the
world,
tempo primo
cresc.

cont'd

Ex.5.13 cont'd

cont'd

Ex.5.13 cont'd

Ex.5.14

Ex.5.15.a

Ex.5.15.b

Ex.5.16

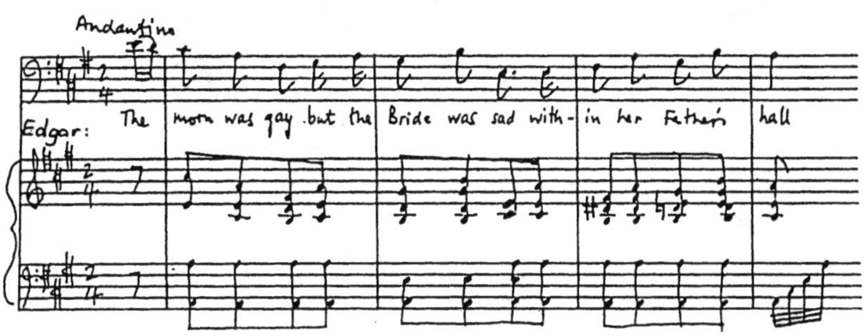

Ex.5.17

Ex.5.18

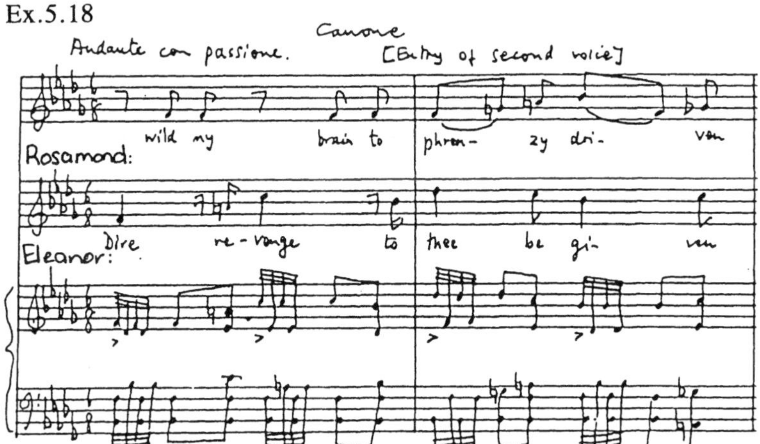

Ex.5.19

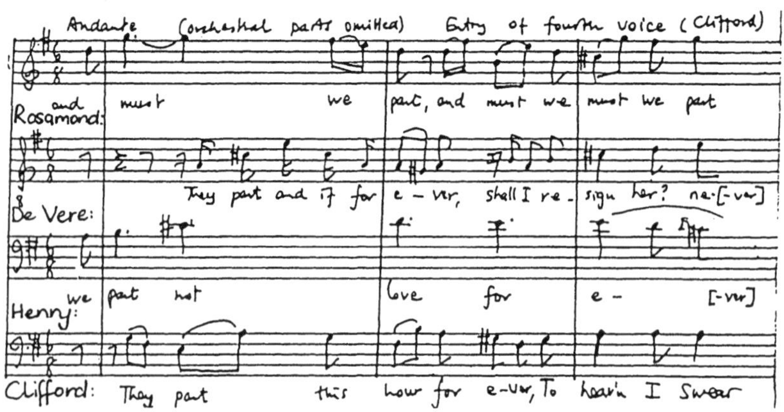

Ex.5.20

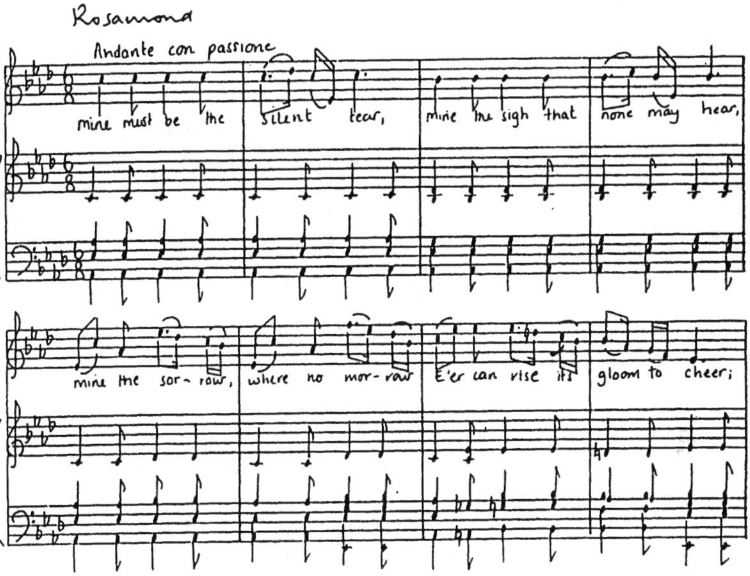
Rosamond
Andante con passione
mine must be the silent tear, mine the sigh that none may hear,
mine the sor- row, where no mor- row E'er can rise its gloom to cheer;

Ex.5.21

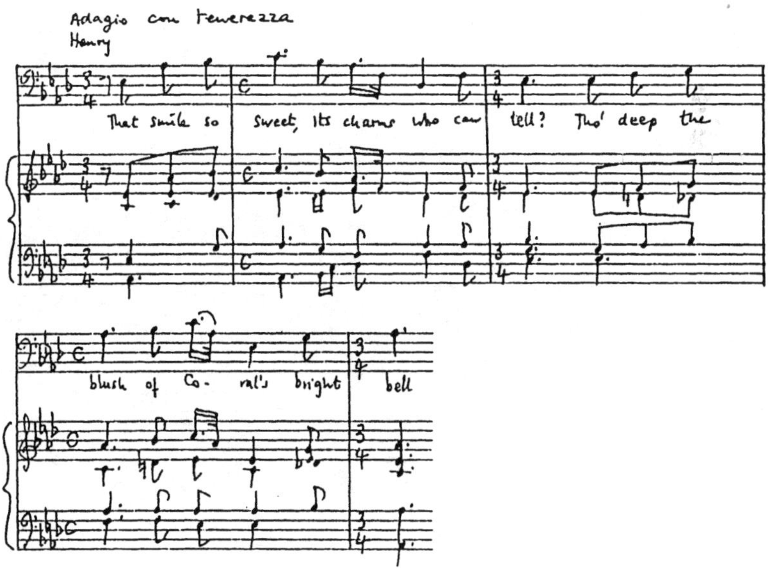
Adagio con tenerezza
Henry
That smile so sweet, Its charms who can tell? Tho' deep the
blush of Co- ral's bright bell

Ex.5.22

Adagio Rosamond:
Oh no no thou'lt mer-cy mer-cy show make me thy slave 'Tis
p Adagio
sad to die in youth to be in the cold cold grave.
pp
cresc.
ff Andantino
Queen
This for thy heart quick quick kneel. Rosamond Oh save me from that mord'rous
Ben marcato
Queen
steel Thy doom is fix'd then take this bowl drink drink

Ex.6.1

Ex.6.2.a

cont'd

Ex.6.2.a cont'd

Ex.6.2.b

cont'd

Ex.6.2.b cont'd

Ex.6.3.a

Allegro Agitato
Giselle
63
rite Yet, strange to say, It was not day, Yet, strange to say, It was not
fp
day, The full Moon shone round and bright...........
poco meno mosso
pp
And save her rays no other light........................ Was beaming on the
cres:
colla voce
shrine, When, as I was about to
a tempo
p
I dreamt we stood.

Ex.6.3.b

Ex.6.4.a

cont'd

Ex.6.4.a cont'd

Ex.6.4.b

Ex.6.5

Ex.6.6.a Ries

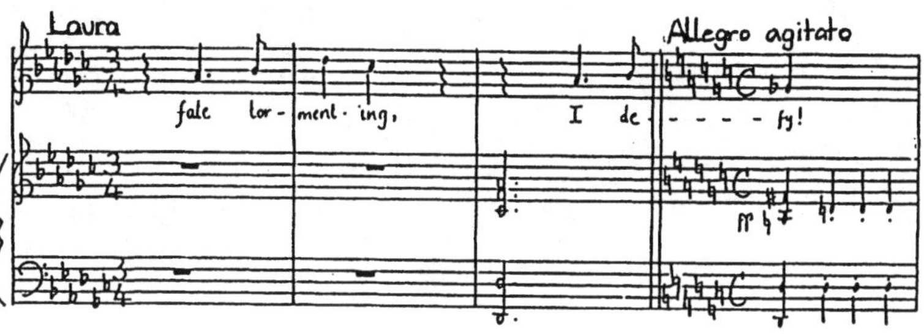

Ex.6.6.b Loder

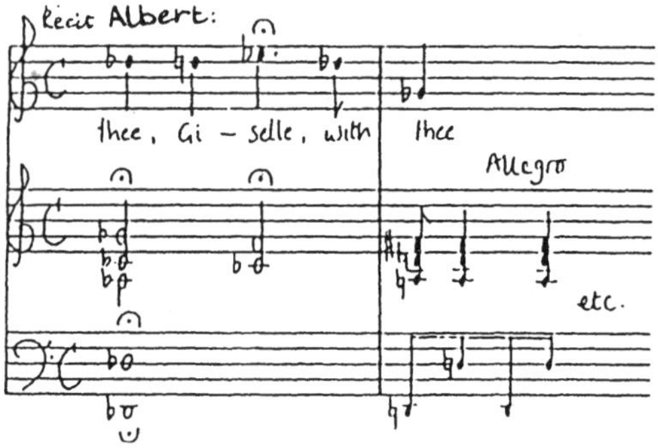

Ex.6.7

cont'd

Ex.6.7 cont'd

Ex.6.8

cont'd

Ex.6.8 cont'd

love all life is stir-- ring In the air the wood the lake... And the
sun flames high oer the mountain Wake dearest wake, oh. wake dearest
wake... dearest wake..... Oh dear------ est a ---wake.... dearest
wake dearest wake...... dear---- est A-wake a--wake a---- wake.

Ex.6.9.a

ALLEGRO.
ff
Duke
For one sweet hour, a_las, too brief! . . . Pride,
p
pomp and pow'r a_side we fling And o'er our woodman's fare we'll
laugh As tho' state were an i......dle thing. For
This song was first entitled "Believe me, honest friends."

Ex.6.9.b

Ex.6.10.a

Ex.6.10.b

Ex.6.11

cont'd

Ex.6.11 cont'd

cont'd

Ex.6.11 cont'd

cont'd

Ex.6.11 cont'd

ad lib.
me. Rouse, love; rouse, love; shake off this coun--ter--feit of
Andante.
death; I feel her bal----- my breath! She sighs— her ru--by
a tempo.
a tempo.
ad lib.
lips unclose; Hush, hush! Horror! that death-denouncing bell Hath
(bell)
slain my love— my life— my bosom's peer--less queen!
a tempo.

Ex.6.12.a

* * See Ex.6.24

** ** See Ex.6.12.b

Ex.6.12.b

Allegro non troppo.
My fa------ther, on his dy-ing bed, Made me
vow you Maid to wed. Ah Heav'n! I found her in a low--ly cot, I
watch'd with care her steps thro' life; I found her in a low--ly cot, I
cres.
watch'd with care her steps thro' life; Whether I love, or love her not, She
must be--come my wife. No! I for--bid such cru-el deed Which Heaven's

Ex.6.13

Theodore
There now! but is there nought else that I can do?
pp
ISABELLINA
Andantino con moto.
Pray just withdraw that curtain, before yon martyred saint.
(he withdraws the cur-
p
pp
- tain.)
ad lib.
All° non troppo.
Tutti
How lovely! And now I look a-gain 'tis strangely like the Lady Agnes! Yes! 'tis
ad lib.
f
Sopi
Ten.&Bassi.
Sup.&Ten.
(Enter H.)
BARON.
like the Lady Agnes, very like, very like! very like, very like very, very, very. Silence!
f
(They all hurry off.)
(exits)
away! be-gone!
What a horrid countenance!
dim.
p

See * *

Ex.6.14.a

ad lib. BARON.
Allegro moderato
Ma..drid, Oh! Ma...drid! I see thee once a....gain!
She whom I tore from out her husband's halls Still is be..fore me!
Fran..tic with despair! A..gain that in...jur'd Lord be..

Ex.6.14.b

Allegro ma non troppo
62
Baron
thrills? No!. It is not so! No! It is not
so!
espress.
Allegro agitato.
It is re--- morse, Which like the blast... Before this
Al---tar Brings back the past: With burn-ing key...
it opes the heart... To let in a--go--ny, a-------go--

Ex.6.15

Allegro moderato
110
Baron
agitato.
-neath my dagger falls! I see his wound-ed breast! his dy--ing
-glare!.... Mem---o---ry! Oh! mem---o---ry of guilt.... And
lento.
pain!
Moderato.
cres.
dim.
Andante sostenuto.
When o---thers at the watch-fire slept In
calm and joy--ous dreams. When o---thers at the watch-fire slept In

cont'd

Ex.6.15 cont'd

111
poco accel.
calm and joy---ous dreams Pale trou_bled forms Around me
wept Or woke me with their screams! . . . . . . The
poor_-est Vas_-sal in my hall Might scoff at all.. my pow'r, The
pow'r of him, whose heart, the pow'r of him whose heart can ne'er re_- call One calm or happy
pp
hour. The pow'r of him whose heart can ne'er re_-call One calm or hap_py

Ex.6.16

Andante sostenuto
112
Baron
hour, one hap-py hour!
pp
Allegro agitato.
No! No!
f fz fz
No cy-press o'er my tomb shall wave, My mem'-ry claim no
ff p crea
tear, The trav'l-ler pas-sing by my grave Will cross himself with
f p
fear Gone for ever hope of
mf

Ex.6.17

Andantino, quasi Allegretto.
p
cres.
f
dim.
In vain, in vain I wan..der thro' ev'..ry dreary cham..ber, In vain I call .. thee, my Raymond, ev'..ry...where . . . . . I call thee ev'..ry ..where . . . . I call thee ev'..ry...where . . . . yes, ev'..ry..where Echo a..

cont'd

Ex.6.17 cont'd

lone re ......spond eth, re..spondeth to my anguish, Echo a..lone re.....
..spond..eth, re..spondeth to my anguish, And thy name, thy dear name but mock
.......eth my des..pair....... mocketh my des..pair..... mocketh my des
..pair.
cres.
dim.
ff
p
cres.
espress.
f
dim.
p

Ex.6.18.a

Ex.6.18.b

Ex.6.19

Lento.
pp
trem.
fz
fz
ff
(Antoni appears
As Antoni withdraws his head from window, Raymond is conducted in.
Andante. 8va
"Ah that way bolts are withdrawn"
p
Andante con moto.
p
RAY.
Fare_well the forest and the plain, Fare___well the blue and star_ry
p
sky; My ear___ly doom, a dungeon chain, To suffer, or perchance to

cont'd

Ex.6.19 cont'd

die. But can a dungeon chain e---rase While mem'--ry lives, my
Ag---nes dear. That an---gel-form, that an---gel-face Thy sigh of Love thy parting
poco rall.
pp
tear..... That an--gel form that an--gel face Thy sigh of Love.... thy
a tempo
pp
part---ing tear!
They'll seek me in my father's home, For me, they'll ring the parting
p

Ex.6.20

Moderato maestoso.
Now in her eyes those trem_bling
tears Enchanting, more... than language prove;... Her blushing cheek, such beauty
wears... But for my crime I still could love! So on the night flow'r
hangs the dew, But with the sun_light, the sun_light melts away. She smiles! my

Ex.6.21

Allegro non troppo
160
Agnes
Hold he is not conscious! No! he
Madelina
Raymond
my father's name! hour of vengeance! not conscious?.
Theodore
Stay, Ferdinand!
sleepeth!
he sleepeth!
he sleepeth!
THEO.
he sleepeth!
pp
Andante.
AGNES.
Lost! and in a dream, His
THEO.
pp
Andante.
Andante. Lost! and in a
p
pp
Raymond & Agnes.

cont'd

Ex.6.21 cont'd

eyes on phantoms beam: Be-hold that look so pale, Dark
dream, His eyes on phantoms beam, Be-hold! Dark
pp
sha-dows him as-sail!
HADE.
Lost! and
sha-dows him as-sail! Lost! and in a dream His
colla voce.
mf dim.
pp
in a dream Behold that look so pale, Dark sha-dows
eyes on phantoms beam Be-hold that look so pale
him as-sail, him as-sail! Lost!
RAY.
Dark sha-dows him as-sail! Lost! and in a
THEO.
Dark sha-dows him as-sail!
SAROH.
Lost! and in a

Ex.6.22

Allegro non troppo
Raymond
122
justice would disown, And who art thou, that I should heed; Be_fore my
ff
dim..
father's will, thy frown!
cres
dim.
fp
Andante espress:
While yet in boyhood's ro_sy morn, A Brigand sought our happy home; My
p
Mo_ther from these arms was torn, My fa_ther met an ear_ly doom. He
sunk beneath the Murd'rers steel, And left in sorrow, past re_lief, A
pp

cont'd

Ex.6.22 cont'd

BARON.
blight_ed heart too crush'd to feel, A_ _ _lone in life, a child... of grief! With
(aside.)
hor_ _ror and with dread I hear this fear_ful, fear_ful his_ _to_ _ry! And
(aloud)
pp
you are of Ma_drid, And this your life's.... dark mys_ _te_ _ry? Pro_ _
_ceed! Pro_ceed!
RAT.
That hour fair Ag_nes'
smile I met, How chang'd my path of lone_li_ness! I felt there was one

Ex.6.23.a

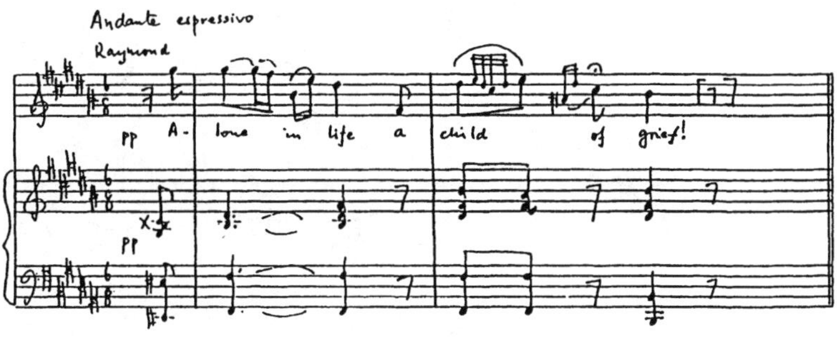

Ex.6.23.b

Ex.6.24

Agnes
39
Andantino.
pp
Oh Agnes! martyr fair, Behold me lowly
kneel....ing Hear thou my hum.ble pray'r Each in.most thought re.
..veal.......ing! May my pray'r Calm my care These sad doubts re..
mf
..pell.....ing, Hear! oh hear! La...dy dear! While my beads I'm
f dim
tell...ing.. Oh Ag...nes! mar..tyr fair, Be..hold

Ex.6.25

Allegro
Agitato.
p
cres.
8va
f
ff
Larghetto sostenuto.
Men, who with relentless hearts, ... Ev'ry
law of ho--nor break; Think! more swift-ly life de--
--parts . . . . . . Than the sun--light . . . . from the lake.

Ex.7.1

Ex.7.2.a

cont'd

Ex.7.2.a cont'd

Ex.7.2.b

Ex.7.2.c

Ex.7.3

Ex.7.4.a

cont'd

Ex.7.4.a cont'd

Ex.7.4.b

Ex.7.5

ALLEGRO AGITATO
VALMOUR.
Why seeks my soul in ev-'ry face that on-ly look of her's to
trace? sweet poi-son flow'r, thy hon - ied breath but once, but once im-bib'd

Ex.7.6.a

Andantino
MICHEL.
When I be-held the an - chor weigh'd, and with the shore thine
im - age fade, I deem'd each wave a bound-less sea that bore me still from love and

Ex.7.6.b

Adina
prendi, per me sei li - be - ro: re - sta nel suol na - ti - o;
Cantabile
non v'ha destin si ri - o, che non si can - gi un dì, resta.

Ex.7.7.a

Marcella.
Thus to - ge - ther, side by side, love, like two ring - doves,
Tempo di Waltz. p
we will stray, we will stray; yes, u - nit - ed.... we through life

Ex.7.7.b

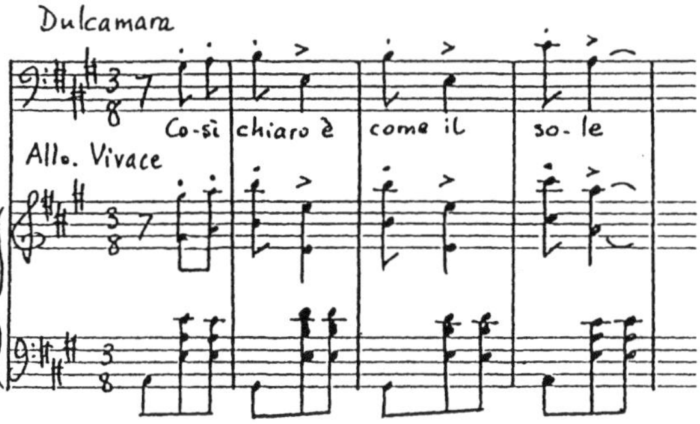

Ex.7.8

cont'd

Ex.7.8 cont'd

Ex.7.9

cont'd

Ex.7.9 cont'd

now a - way! where those bow'rs of myr-tle wreath in that maze, that maze of
stream.... and tree, ... beats there still a heart to breathe one sad
sigh for me, one sad sigh, one sad sigh for me?
ad lib. e lento.
one sad sigh, one sad sigh.................... for me?

Ex.7.10.a

Ex.7.10.b

Ex.7.11

Ex.7.12

Andante mosso Elizabeth
almost spoken
My re-solu-tion may seem some-what strange

Ex.7.13

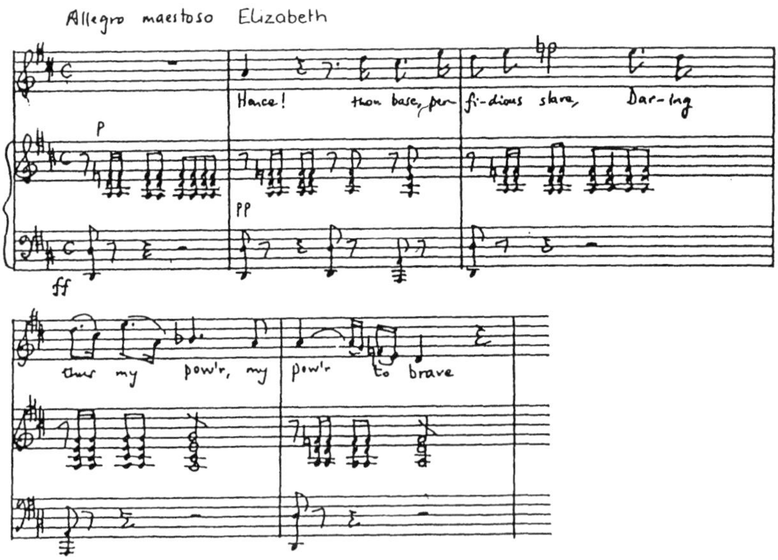
Allegro maestoso Elizabeth
Hence! thou base, per-fi-dious slave, Dar-ing thus my pow'r, my pow'r to brave

Ex.7.14.a

Ex.7.14.b

Ex.7.15.a

Ex.7.15.b

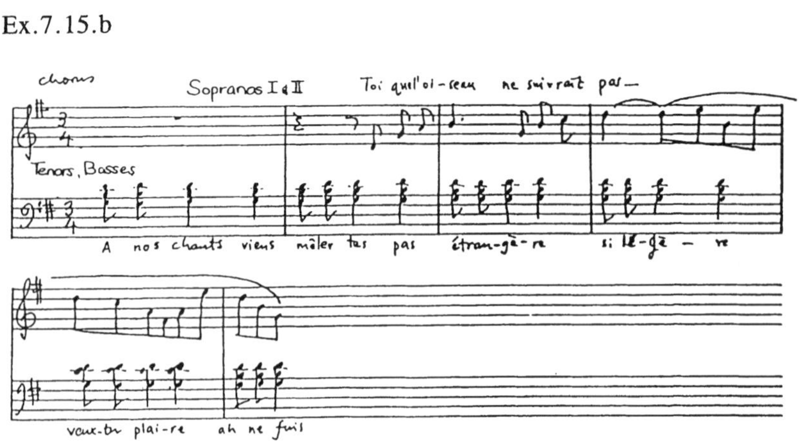

Ex.7.16.a

Balfe

Ex.7.16.b

Liszt

Ex.7.17.a

cont'd

Ex.7.17.a cont'd

Ex.7.17.b

When other lips and other hearts their tales of love their tales of love shall tell in
[sic]
[sic]
[sic]
language whose excess imparts the pow'r they feel the pow'r they feel so well there
[sic]
may perchance in such a scene some re[-] col[-] lec[-] tion be of
[sic]
days that have as happy been as happy been and you'll remember
[sic]
me
[sic]

Ex.7.17.c

Andante
when other lips and other hearts their tales of love shall tell
[D?]
[Eb?]
[?♮]

Ex.7.17.d

Cantabile sostenuto
Amina
a tempo
[a-]mor del mio, del mio di- let-to a-
a tempo
mor, a-mor la co-lo-rò, a-
f
p
mor, a- mor del mio di- letto.

Ex.7.18.a

cont'd

Ex.7.18.a cont'd

Ex.7.18.b

Andantino
SARA.
So - gna - i che fra ca - te - - ne D'am - bi - ta ser - vi - tù A me tra - es - se I - me - Fio - ri - ta giò - ven - tù Un Rè d'un gran - de im - pe - -
Arpa.
Corno.
Vllo
Fl.
Cl.
Quat.
pp Sourd.
pp
S.

Ex.7.19.a.

THADDEUS.
Allegro maestoso.
With-out
friends, and with-out a home, my coun-try, too! yes, I'm ex - il'd from thee; what fate, what
fate a - waits me here, now pi - ty, Heav'n! oh calm my de - spair!
lento.
colla voce.
dolce assai.
Moderato.
rall.
'Tis sad to leave our Fa - ther-land, and friends we there lov'd

Ex.7.19.b

Andante cantabile in Eb [clearly added later]
Tis sad to leave yr [?] fath-er-land and friends you lord [sic] there well to
wan-der on a stranger strand where friends but sel-dom dwell yet hard as are such
ills to bear and [sic] deep-ly tho they smart their [sic] pangs are light to
[sic]
[notes crossed out]
those who are the or-phans of the heart the or-phans the
or- phans phans of [superimposed]
[sic]
the heart
[sic]

Ex.7.19.c

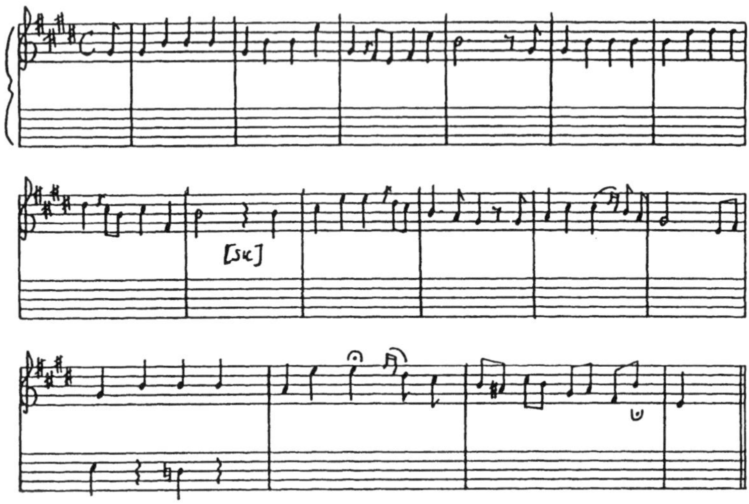

Ex.7.20

cont'd

Ex.7.20 cont'd

eye to weep when I might grieve, one bo-som to re-ceive the sigh which
sor-row oft will heave, one heart the ways of life to cheer, tho' rug-ged, tho' rug - ged they might
pp
be, no language, no lan - guage can express how dear that
f

Ex.7.21

Allegro marziale grandioso.
Thaddeus.
When the fair land of Po - land was plough'd by the hoof of the
ruth - less in - va - der, when might with steel to the bo - som, and

Ex.7.22.a

Ex.7.22.b

Ex.7.23

Ex.7.24

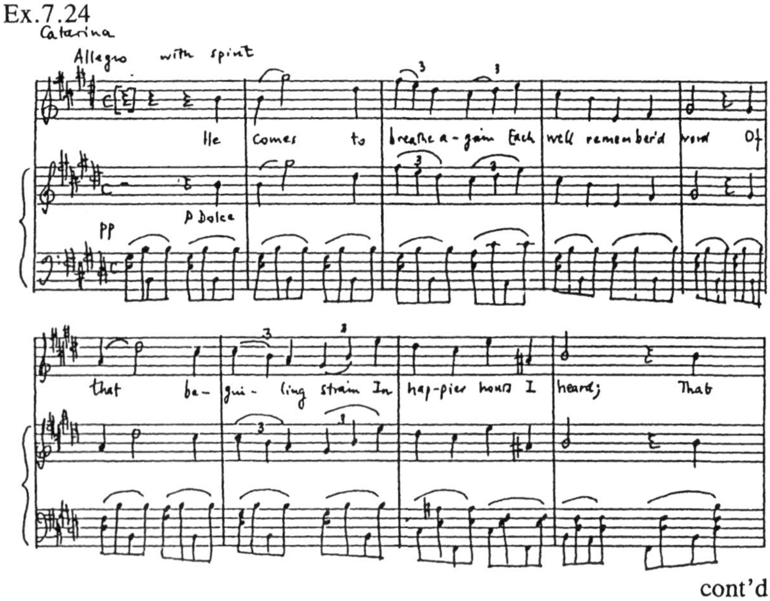

cont'd

Ex.7.24 cont'd

Ex.7.25.a

Ex.7.25.b

Ex.7.26.a

Ex.7.26.b

MORENIGO
All° non troppo.
f p
pp
-roil De Chy - pre le peu - ple cou.
-pable Exila de ses rois l'il_lustre descen_dant, Et Veni - se,toujours au malheur- secou.
-rable, A ju-ré de s'u - nir au dernier Lusi - gnan. De ce pros.
ANDRÉA.
Eh bien!
-crit nous re-levons le trô - ne, Et pour que rien ne brise dé-sor -

Ex.7.27

The task, how-ever sad or stern, 'tis mine to speak, Thine is the sad propos'd to
learn The birth to seek lis- ten.

Ex.7.28

[Larghetto] Adolphe (with great feeling)
pp
With- out re- proach with-out an
an - gry thought which just - ly might, which justly might a-rise

Ex.7.29.a

Larghetto cantabile
[sic]
dear the ear-ly dawn-ing
un-to [?*]-ous eyes
when the beams of morn-ing
O'er the waves arise
*indecipherable

Ex.7.29.b

cont'd

Ex.7.29.b cont'd

Ex.7.30

Andante.
tremolo.
pp
f
1mi e 2di Soprani.
1mi e 2di Tenori.
Bassi.
Up - ward from the
Up - ward from the
Up - ward from the
ne - ther world, the ne - ther world, thro' the deeps of end - less space, of end - less
ne - ther world, the ne - ther world, thro' the deeps of end - less space, of end - less
ne - ther world, the ne - ther world, thro' the deeps of end - less space, of end - less

Ex.7.31

Arimanes
Allegro moderato
Since he has dar'd to brave, he shall be-come my slave,
let then that task be thine, en-snare, and make him mine, mine!
since he has dar'd to brave, he shall be-come my
slave, let then that task be thine, make him
cresc.

cont'd

Ex.7.31 cont'd

cont'd

Ex.7.31 cont'd

page; as such with him en - gage, as such with him en -
pp
poco riten.
a tempo.
gage! Mark well, mark well this man, ensnare, and make him
riten.
mine, be thou his page, en - snare, and make him mine,
tempo.
mine, mine! .... since he has dar'd to brave,
tempo.
p

Ex.7.32

Andantino.
dolce.
Clarinetti.
Violoncelli.
cresc.
Rupert.
No prize can fate on man be - stow like love in wo - man's
breast; a light that bright - est shines in woe, and bless - ing then in

cont'd

Ex.7.32 cont'd

cont'd

Ex.7.32 cont'd

cont'd

Ex.7.32 cont'd

poco riten.
rall.
me still live, where falls be - low the soft, ... the soft warm light of
pp
accel.
cresc.
love! yes, leave me on - ly love! the soft warm light of love!........... let me still
p poco accel.
cresc.
cresc.
live where falls be - low the light of love, yes, leave me on - ly love! the soft, warm light of
dim.
p
cresc.
love!................... let me still live where falls be - low the light of
riten.
f
tempo.
love!

Ex.7.33

Allegro ma non troppo
Duchess
pas-sions black and mur-ky night veils fair pi-ty
from my sight
etc
'Tis re-venge 'Tis re-venge, re-venge, re-venge,
'tis re-venge re-venge re-venge whose lur-id hand
Stacc:
p cresc.
f
pp
cresc.
ff
Accel.

Ex.7.34

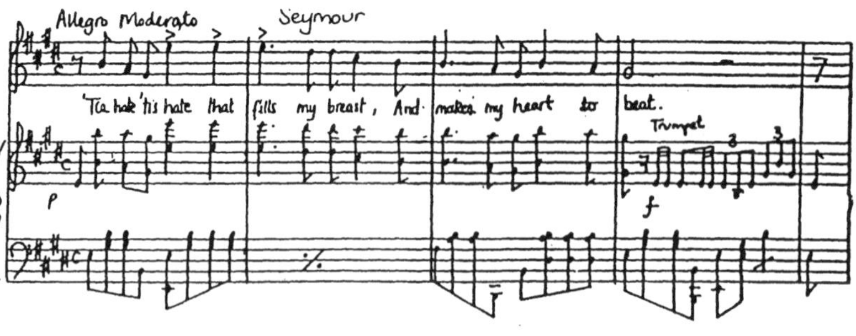
Allegro Moderato
Seymour
'Tis hate 'tis hate that fills my breast, And makes my heart to beat.
Trumpet
p
f

Ex.7.35.a

La Spagnuola
Moderato Allegretto e Grandioso
Blanche
As the mountain streams when the bright sun-beams
when the bright sun beams Their surface fleck and so
gai-ly deck the di-a-mond spray in co-lours gay.
pp
Cresc.
p

Ex.7.35.b

Andantino
Marie
pp
As you hope to mer-it mercy from a-bove;
Ne'er be-lieve I fos-ter'd this man's guil-ty love.
Though my girl-ish fol-ly triumph'd for a while
Riten.
Guilt dark guilt my bo-som
Tempo
Ne'er did yet de-file
col canto
cresc.
pp

Ex.7.36

Ex.7.37

cont'd

Ex.7.37 cont'd

poco più animato
and flow - ret
tell.
Our
glances left this dull cold earth
cresc.
pp
fp
To
gaze on heav'n a-bove, on heav'n a-bove: p Ah!
we
Tempo primo
felt 'twas there our thoughts had birth
cresc.
we felt 'twas there, 'twas there came
f
Riten:
Adagio
love - - - -
we felt 'twas there came love!
Stentato
fp riten: col canto
etc.

Ex.7.38

Mane
Larghetto quasi adagio
Corni.
poco riten.
Poco riten.
a Tempo
Ritan.
a piacere
Slentate
Col Canto: cresc.
Colla parte
a Tempo
Riten.
Adagio assai

Ex.8.1

SCHERZANDO.
MARITANA.
It was a Knight of princely mien,
Pel bos- - - - co, fuor dal suo cas-tel,
Ped.
p
One blue and gold-en day
D'un gior-no sul ca-der,
leggiero.

Ex.8.2.a

Lento e molto cantabile
319
Don Jose
joy's last hope for- - sa - - - - ken, Sinks in des-pair, lost in despair, in deep des-
non po - trei... lo sen - - - to La vita, ohimè, la vita ohimè, più sop - por-
rall.
-pair.
-tar.

Ex.8.2.b

Rosina
e cen-to trap- po - le prima di ce- de- re fa-rò gio-
Moderato
car, fa- rò gio- car,

Ex.8.3.a

Lento
Maritana
d him, ... And thy gentlest, and . . . thy gentlest pi - - - ty
Fa . . . . che ar - ri - da, Fa . . . che arri - da a . . . .
Lazarillo
d him, ... And thy gentlest, and . . . thy gentlest pi - - - ty
Fa . . . . che ar - ri - da, Fa . . . che arri - da' a . . . .

lend him
lu - i,
lu - i,

Ex.8.3.b

Ex.8.4

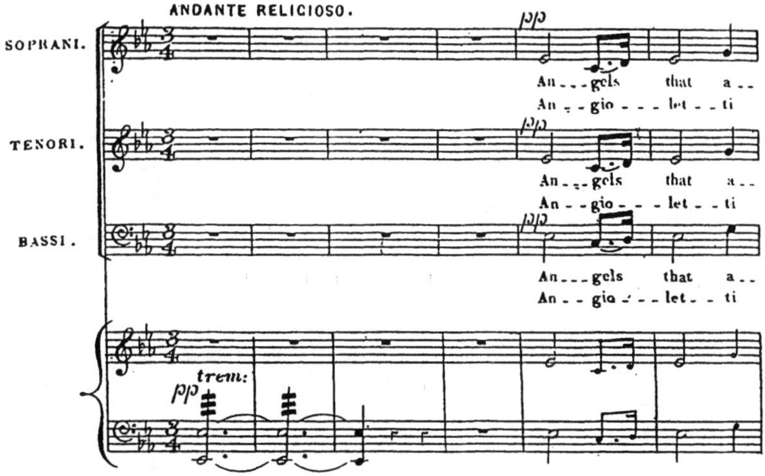

cont'd

Ex.8.4 cont'd

- - round us hover, Guard us till the close of day,
che nel cie - lo Fiam - meg - gia - - te di splen - dor,
- - round us ho - ver, Guard us till the close of day,
che nel cie - lo Fiam - meg - gia - - te di splen - dor,
- - round us ho - ver, Guard us till the close of day,
che nel cie - lo Fiam - meg - gia - - te di splen - dor,

Ex.8.5

ALLEGRO MODERATO.
MARITANA.
Of Fai_ry wand had I...... com_mand
Se a_ves_si l'ar_ _ _ te d'in_ _ _can_tar,

Ex.8.6

CANTABILE E CON MOLTA SEMPLICITA.
8va
pp
Ped:
Maritana
Scenes that are bright - - est, May charm.... a - - while.....
Fes - - - te bril - - lan - - - ti, Tap - - pe - - - - - ti e fior,.......
Hearts which are light - - est And eyes.... that smile:
Per - po - chi is - tan - - ti Fan lie - - - - to il cor:

Ex.8.7

Allegro molto
Cres:
earth.... to share Thy hap-pi-ness or woe: Yes! on earth to
Cres.
share.................. .... to share, thy hap-pi-ness thy hap---- pi-
p
ff
ness........ or woe
stringen:

Ex.8.8

ADAGIO MA NON TROPPO.
4 Corni.
Fl:
Ob:
Clar:
Fag:
Ped:
Bassi.
Clar:
Ob:
Corno.
Fag:
Fl:
Ob:
soave.
Clar:
Fag:
Corno.
8va

Ex.8.9

Ex.8.10

cont'd

Ex.8.10 cont'd

Ex.8.11

Ex.8.12

Andante
Poco Più Lento.
con dolore.
Celli.
LURLINE.
dolente.
rall;
In Tempo.
Ah no! on me your rage, on me your rage let fall; For-
Clar: & Fag:
agitandosi.
-give-ness I im-plore, On me your rage let fall. For-give-ness, for-
Ob:
Fl:
give-ness, for-give-ness I im-plore, I im-plore;
Soave.
Him to his
Viol:
Celli
father's halls, Me to thine arms restore. His life in mercy
Oboe.

cont'd

Ex.8.12 cont'd

Ex.8.13.a

cont'd

Ex.8.13.a cont'd

Ex.8.13.b

Ex.8.14

Ex.8.15

cont'd

Ex.8.15 cont'd

Ex.8.16

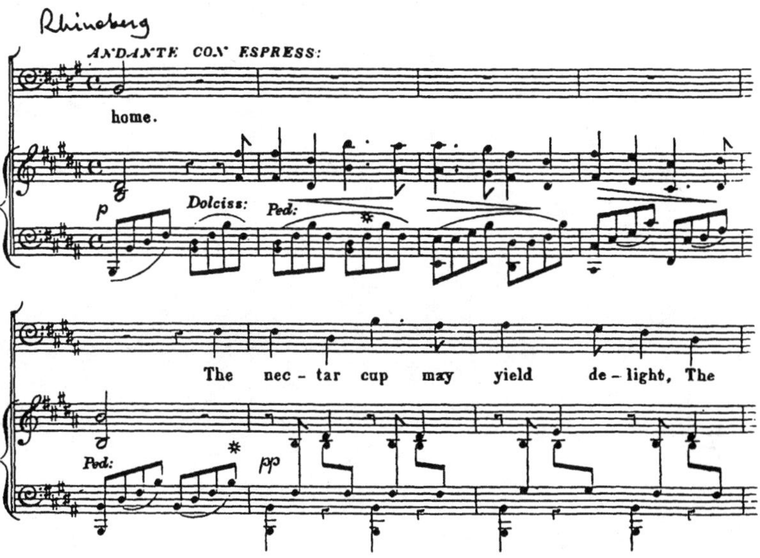

cont'd

Ex.8.16 cont'd

Ex.8.17

Ex.8.18

Allegro moderato

Ex.8.19

ANDANTINO.
Con grazia.
Princess
Carelessly cantering
down a lone_ly dell!
Where thro' the tan_gled branches scarce a sunbeam fell,

Ex.9.1.a

Ex.9.1.b

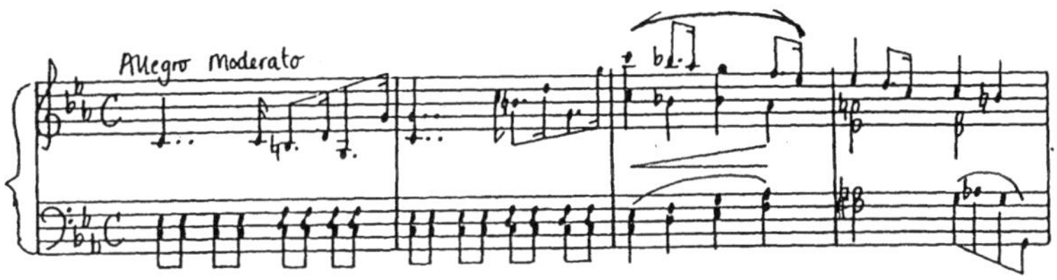

Ex.9.2

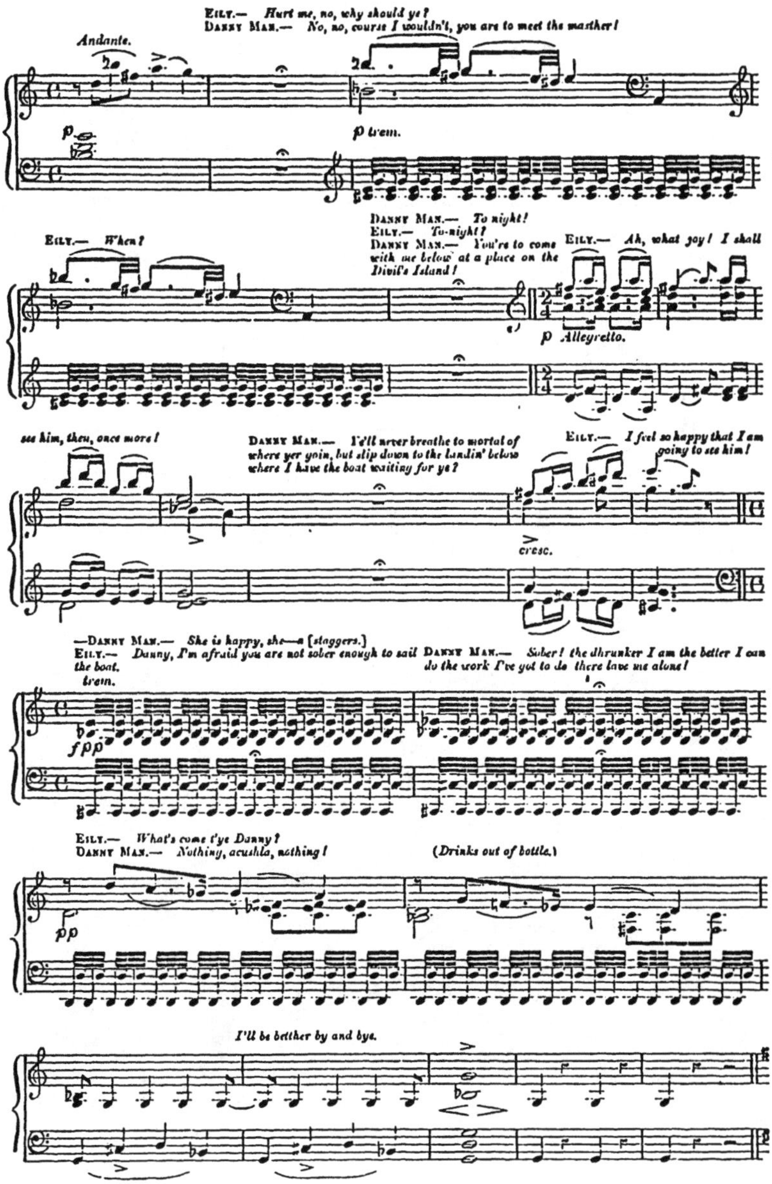
Eily.— Hurt me, no, why should ye?
Danny Man.— No, no, course I wouldn't, you are to meet the masther!
Andante.
p
p trem.
Eily.— When?
Danny Man.— To night!
Eily.— To-night?
Danny Man.— You're to come with me below at a place on the Divil's Island!
Eily.— Ah, what joy! I shall see him, then, once more!
p Allegretto.
Danny Man.— I'll never breathe to mortal of where yer goin, but slip down to the landin' below where I have the boat waiting for ye?
Eily.— I feel so happy that I am going to see him!
cresc.
—Danny Man.— She is happy, she—a (staggers.)
Eily.— Danny, I'm afraid you are not sober enough to sail the boat.
Danny Man.— Sober! the dhrunker I am the better I can do the work I've got to do there leave me alone!
trem.
fpp
Eily.— What's come t'ye Danny?
Danny Man.— Nothing, acushla, nothing!
(Drinks out of bottle.)
pp
I'll be betther by and bye.

Ex.9.3

Andantino con moto.
mf
Myles.
Let the far-mer praise his grounds, let the hunts-man praise his hounds, the
shep-herd his dew-scent-ed lawn— but I more bless'd than they spend each
hap-py night and day with my charm-ing lit-tle cruis-keen
lan, lan, lan,.... my charm-ing lit-tle cruis-keen lan!
p
cresc.

Ex.9.4

RECITATIVE.
Assai moderato.
dolce.
p
MYLES.
From In-chi-ge-la, all the way I
pp
tra-vell'd un-to Ker-ry, and migh-ty wea-ry seem'd the way, my
poor heart was not mer-ry:
QUASI RECIT.
to ev'-ry ca-bin door there
pp

Ex.9.5

Allegretto. leggiero.
con grazia.
cresc.
MYLES.
leggierissimo.
pp
It is a charm - ing girl I love, she
comes from Gar - ry o - wen; she's gen - tler than the tur - tle dove, her
hair is brown and flow - ing! her eye is of the soft - est blue, her
cresc.

Ex.9.6

Andantino.
EILY.
In my wild mountain val-ley he
sought me, my heart soon he knew was his own; when he made me his bride then he
taught me con-tent-ed to dwell here a-lone! when the day in the west is de-
-clin-ing, his boat on the dark lake I see, ... and led by my ta-per's bright
cresc.
dim.
con grazia.

Ex.9.7

Andante mesto.
dolce.
cresc.
Eily.
dim.
ppp
I'm a - lone, I'm a - lone,.. I watch the
stars.... as they rise, I hear the sound of my sighs.. mock'd
by the breez - es' moan! all things round me seem to say that I am
sempre pianissimo.
sad and so are they, so.... are they! but could I see my heart's de-light, his

Ex.9.8

Andante con moto.
leggiero.
f
Ann Chute.
The eye of love is keen, the
p
eye of love is keen, and rea - di - ly can trace
in the lov'd one's face...... the pass - ing shade, the
ritenuto.
pass - ing shade that to the world re - mains un - seen! the
ritenuto.
f

Ex.9.9

Andante espressivo.
f
dim.
HARDRESS.
p
Ei - ly Ma-vour - neen, I see thee be-fore me,
fair - er than ev - er with death's pal - lid hue;....
mor - tal thou art not, I
hum - bly a-dore thee,
cresc.
yea, with a love which thou know - est is true!
cresc.
dolce.
look'st thou in an - ger, ah, no, such a feel - ing
p

Ex.9.10.a

Allegro moderato assai.
DANNY MAN.
Trust me, trust me, that glove, that
glove would be a to - ken as plain as
a - ny word that's spo - ken, that glove would be a
to - ken as plain as a - ny word that's spo - ken!
MRS. CREGAN
That

Ex.9.10.b

Moderato
Lucia
ti rim- pro- ve- ro ta-
pp
cen - il mio stra - zio
il mio do- lo- re.

Ex.9.11

Allegro marziale e maestoso
Mrs Cregan.
No blot on our scut-cheon shall e'er have a place, but
pure as of old shall be Cregan's high race; dis-hon-our has threaten'd but threaten'd in
cresc.
vain, the tow'rs of Tore Cre - - gan rise proud - ly a -
- gain, the tow'rs of Tore Cre - gan rise proud - ly a -
- gain! Bad luck to the eyes and the Col-leen's fair face, that
Danny Man.
p
cresc.

Ex.9.12

DANNY
Andante espressivo.
cantabile.
The Col - leen Bawn, the
Col - leen Bawn from child - hood I have known, I've seen that beau - ty
in the dawn which now so bright has grown; al - though her cheek is
blanch'd with care her smile dif - fu - ses joy, Heav'n form'd in her a

Ex.9.13

DANNY
Allegro con fuoco.
RECIT.
nought, nought shall check me!
Allegro.
RECIT.
well I know my course!
Allegro energico moderato.
cresc.
Du - ty, yes, I'll do my
du - ty,
du - ty, yes, I'll do my du - ty!
8va
what is love and what is beau - ty to a

Ex.9.14

DANNY
ALLEGRO ENERGICO MODERATO
Man, no o - ther law has Dan - na Man, has
Dan - - - - - - - - - ny Man, has
Dan - - - - ny Man; I'll help the mas - ter
where I can, no o - ther law has Dan - ny

Ex.9.15

Allegretto
p
p
cre - - - scen - - - do.
EILY.— What place is this you have brought me to, Danny?
ff dim.
EILY.— It is like a tomb.
DANNY MAN. RECIT.
RECIT.
Step out on this rock,
come, come now, be quick,
Allegro
f
tempo.
RECIT.
the boat is leaking!
andante.
f trem.
dim.

Ex.9.16

[Allegro non troppo – string. il tempo]
(Pushes her off—she clings to the rock.)
EILY.
Spare me for Hardress' sake a-lone!
DANNY MAN.
He wants you dead and gone!
trem.
pp
(Pushes her in—a shot is fired, and he falls in the water.)
ff
fff
dim.
MYLES.
Allegretto.
There is but one Col-leen Bawn, and she does not love
Allegretto.
dolce.

Ex.10.1.a Macfarren: *The Devil's Opera*, overture

Ex.10.1.b Schumann: *Second Symphony*, finale

Ex.10.2.a

Ex.10.2.b

Ex.10.3

Ex.10.4

Ex.10.5

Ex. 10.6

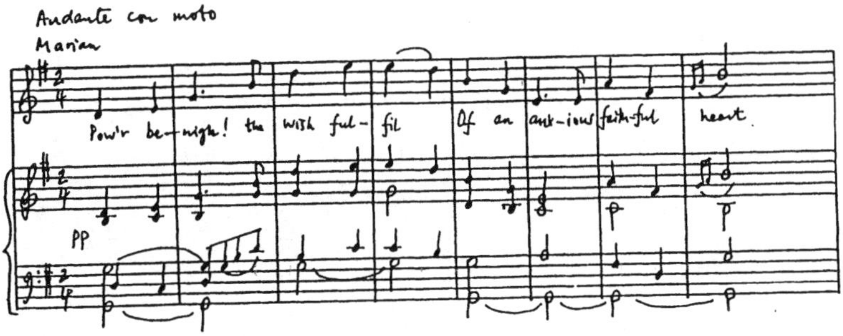

Ex. 10.7

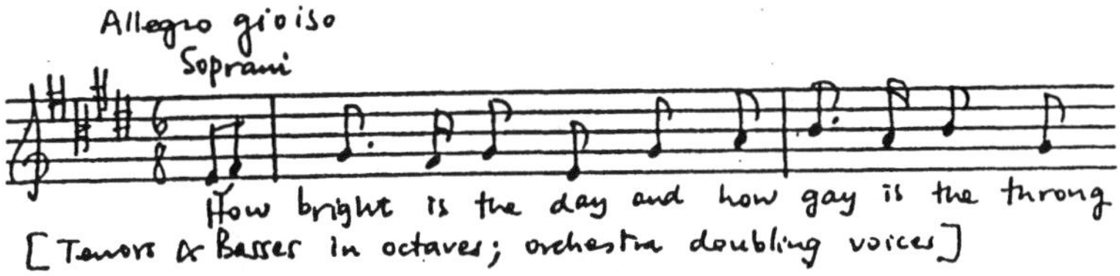

Ex. 10.8

Ex. 10.9

Ex.10.10

Ex.10.11

Ex.10.12

Ex. 10.13

Ex. 10.14

Ex.10.15

Ex.10.16

Ex.10.17

Ex.10.18

Ex.10.19

Andante maestoso.
Squire.
Piano.
f
p
Once a --- gain my fa ----- ther's
hall, Will beam with all . . . . . its an - cient fire; And dear old

Ex.10.20

[Andante soave]
BALLAD.
mf
p
Constance
Why sad_ly
sighs the ev'ning gale Why do you weep pale trembling rose?

Ex.10.21

Allegro giovale MARLOW
Drink friends drink the sparkling bumper drain Like liquid gold the ale behold it
p
cres.
f
cures all pain.
ff

Ex.10.22

(Curtain rises)
PRESTO.
ff
sf
sf
(Enter MARLOW hurriedly)
MARLOW. Would a circumstance so extraordinary be believed! not knowing what to do with the jewels given me in charge by Hastings, that I should ask the churlish old Landlord to place them under lock and key, that they should turn out to belong to said Landlord, that said Landlord should turn out to be no Landlord, but old Hardcastle himself, and this house no Inn after all! What a blockhead have I rendered myself! However, as I could not possibly endure a second interview with Miss Hardcastle, I'll offer no further explanation with her father, but set out for London immediately.
Stay, who can that be I mistook for the barmaid! I cannot possibly quit the hall without taking leave of her, although the difference of our birth, fortune and education makes any connection with her impossible.
Confound it, I must be in love, and with that girl, if ever man was in love, desperately.
attacca.
8557

Ex.10.23

Ex.10.24.a

Ex.10.24.b

Ex.10.25

Ex.10.26

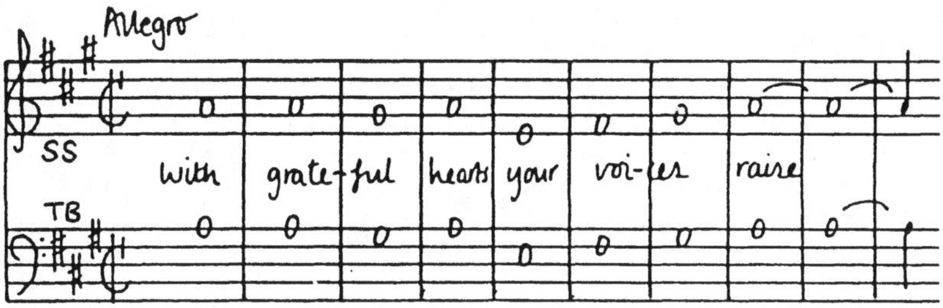

Ex.10.27

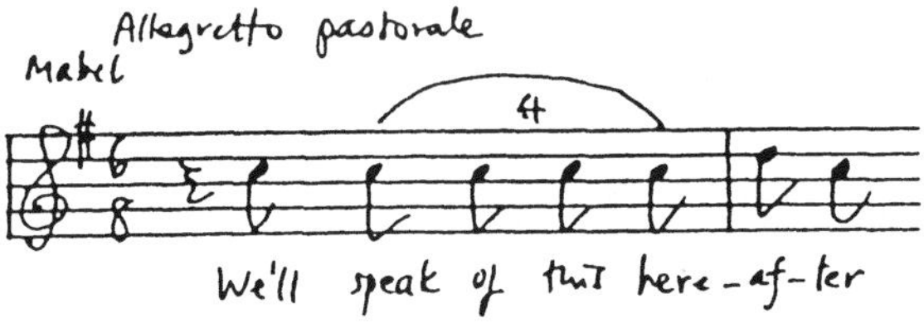

Ex.10.28.a

Ex.10.28.b

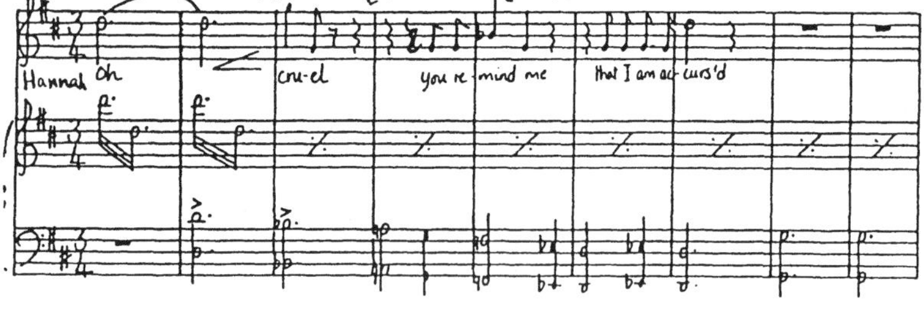

Ex.10.29

Allegro brillante
Hannah
re -
joice
pp
fp
p
f
cres.
sf
re -
joice
ff

Ex.10.30.a

Ex.10.30.b

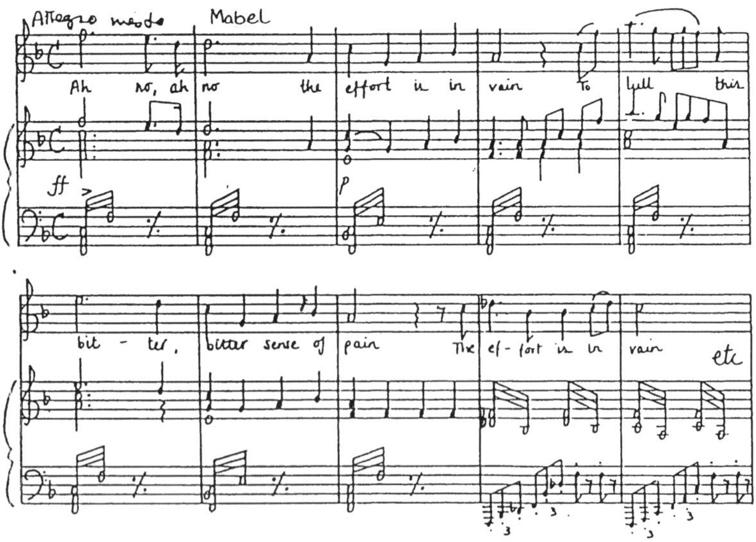

List of English Operas first performed between 1834 and 1864

For the sake of completeness the following list includes Balfe's Italian and French operas and their English versions. These are marked by an asterisk.

All first performances took place in London unless otherwise indicated.

Abbreviations:

?	name unknown
Amphi	Amphitheatre
CG	Covent Garden
DL	Drury Lane
EOH	English National Opera House
GI	Royal Gallery of Illustration
Hay	Haymarket
HM	Her Majesty's
Opéra	Opéra, Paris
Op. Com.	Opéra Comique, Paris
Prn	Princess's
StJ	St James's
Sur	Surrey
TR	Theatre Royal

YR	TITLE	COMPOSER/ LIBRETTIST	THEATRE	DATE
1834	*Nourjahad*	Loder/Arnold	EOH	21.07
	The Dragon	Lee/Morton	EOH	04.08
	The Mountain Sylph	Barnett/ T.J. Thackeray	EOH	25.08
	Hermann, or, The Broken Spear	Thomson/?	EOH	27.10
	Genevieve, or, The Maid of Switzerland	Macfarren/ Baron-Wilson	EOH	03.11
	The Red Mask, or, The The Council of Three	Marliani/Planché	DL	15.11
	The Lord of the Isles, or, The Gathering of the Clans	Rodwell/Fitzball	Sur	20.11
1835	*Sadak and Kalasrade, or, The Waters of Oblivion*	Packer/Mitford	EOH	20.04
	The Shadow on the Wall	Thomson/Serle	EOH	24.04
	The Spirit of the Bell	Rodwell/Kenney	EOH	08.06
	The Covenanters	Loder/Dibdin	EOH	10.08
	The Dice of Death	Loder/Oxenford	EOH	14.09
	Paul Clifford	Rodwell/Fitzball	CG	28.10
	The Siege of Rochelle	Balfe/Fitzball	DL	29.10
	Agnes Sorel	Mary à Beckett/ Gilbert à Beckett	StJ	14.12
1836	*The Rose of Alhambra*	De Pinna/Polhill	CG	12.05
	The Maid of Artois	Balfe/Bunn	DL	26.05
	The Sexton of Cologne	Rodwell/Fitzball	CG	15.06
	The Pacha's Bridal	Romer/Lemon	EOH	08.09
	The Gaberlunzie Man	?/Rede	EOH	26.09
	The Village Coquettes	Hullah/Dickens	StJ	06.12
1837	*Fair Rosamond*	Barnett/ C.Z.Barnett	DL	28.02
	Catherine Grey	Balfe/Linley	DL	27.05
	The Afrancesado, or, Secrecy and Truth	Lee/Serle	CG	19.10

YR	TITLE	COMPOSER/ LIBRETTIST	THEATRE	DATE
1837	*The Barbers of Bassora*	Hullah/Morton	CG	11.11
	Joan of Arc	Balfe/Fitzball	DL	30.11
	Amilie, or, The Love Test	Rooke/Haines	CG	02.12
1838	*The Gipsy's Warning*	Benedict/Linley & Peake	DL	19.04
	The Outpost	Hullah/?	CG	17.05
	Diadesté, or, the Veiled Lady	Balfe/Fitzball	DL	17.05
	Falstaff	Balfe/Maggioni	HM	19.07
	Il Torneo	Lord Burgersh/?	StJ	20.07
	The Devil's Opera	Macfarren/ Macfarren senior	EOH	13.08
	The Foresters, or, Twenty-five Years Since	Loder/Serle	CG	19.10
	Francis I	Loder/Mckinlan	DL	06.11
1839	*Farinelli*	Barnett/ C.Z. Barnett	DL	08.02
	Henrique	Rooke/Haines	CG	02.05
1840	*Lundy in the Olden Time*	Bryan/ Miss Bryan	TR Bristol	27.03
	Fridolin	Romer/Lemon	Prince's (StJ)	26.11
1841	*Keolanthe, or, The Unearthly Bride*	Balfe/Fitzball	EOH	09.03
	The Deer Stalker, or, The Outlaw's Daughter	Loder/Lemon	EOH	12.04

YR	TITLE	COMPOSER/ LIBRETTIST	THEATRE	DATE
1841	*Arcadia, or, The Shepherd and the Shepherdess*	Harroway/ Blanchard	Grecian	19.04
	Il Paddy Whack in Italia	Lover/composer	EOH	22.04
1842	*The Students of Bonn*	Rodwell/ composer	EOH	21.03
1843	*The Queen of the Thames, or, the Anglers*	Hatton/Fitzball	DL	25.02
	*Le puits d'amour**	Balfe/Scribe & de Leuwen	Op. Com.	20.04
	*Geraldine, or, The Lovers Well**	Balfe/trans. à Beckett	Prn	14.08
	The Bohemian Girl	Balfe/Bunn	DL	27.11
1844	*The Brides of Venice*	Benedict/Bunn & Benedict	DL	22.04
	*Les quatre fils Aymon**	Balfe/de Leuven & Brunswick	Op. Com.	15.07
	*The Castle of Aymon, or, The Four Brothers**	Balfe/trans. à Beckett	Prn	20.11
	The Daughter of St Mark	Balfe/Bunn	DL	27.11
1845	*The Enchantress*	Balfe/Bunn	DL	14.05
	The Fairy Oak	Forbes/Coape	DL	18.10
	Maritana	Wallace/Fitzball	DL	15.11
	L'Étoile de Séville	Balfe/Lucas	Opéra	17.12
1846	*An Adventure of Don Quixote*	Macfarren/ Macfarren senior	DL	03.02

YR	TITLE	COMPOSER/ LIBRETTIST	THEATRE	DATE
1846	*The Crusaders*	Benedict/Bunn & St Georges	DL	26.02
	The Night Dancers	Loder/Soane	Prn	28.10
	Loretta, a Tale of Seville	Lavenu/Bunn	DL	09.11
	The Bondman	Balfe/Bunn	DL	11.12
	The Seven Maids of Munich, or, The Ghost's Tower	Rodwell/ composer	Prn	19.12
1847	*The Rival Clans*	Deval/composer	TR Newcastle Upon Tyne	27.01
	Matilda of Hungary	Wallace/Bunn	DL	22.02
	The Forest Maiden and the Moorish Page	Tully/Fitzball	Sur	31.05
	The Maid of Honor	Balfe/Fitzball	DL	20.12
1848	*Jeannette and Jeannot*	Tully/Horncastle Brighton	TR	14.09
	Robin Goodfellow, or, The Frolics of Puck	Loder/composer	Prn	06.12
	Quentin Durward	Laurent/Fitzball	CG	06.12
1849	*Francesca Doria*	Linley/composer	Prn	03.03
	King Charles II	Macfarren/Ryan	Prn	27.10
1850	-	-	-	-
1851	*The Queen of a Day*	Fitzwilliam/?	Hay	13.08
1852	*The Sicilian Bride*	Balfe/Bunn	DL	06.03
	The Devil's In It	Balfe/Bunn	Sur	26.07
1853	*Love's Alarms*	Fitzwilliam/ composer	Hay	17.11

YR	TITLE	COMPOSER/ LIBRETTIST	THEATRE	DATE
1854	*Pittore e Duca**	Balfe/Piave	Teatro Comunale, Trieste	21.11
	*Moro, or, The Painter of Antwerp**	Balfe/trans. Barrett	HM	28.01 1882
1855	*The Minstrel's Return, or, The Sultan's Jubilee*	Rolfe/composer	TR Newcastle Upon Tyne	22.03
	Berta, or, The Gnome of Hartsberg	Henry Smart/ Hawley Smart	Hay	29.05
	Raymond and Agnes	Loder/Fitzball	TR Manchester	14.08
1856	-	-	-	-
1857	*The Rose of Castille*	Balfe/Harris & Falconer	Lyceum	29.10
1858	*Satanella, or, The Power of Love*	Balfe/Harris & Falconer	CG	20.12
1859	*Zaida, or, The Pearl of Granada*	Lutz/Summers	Amphi Liverpool	14.02
	Never Judge by Appearances	Loder/Drayton	Adelphi	07.07
	Victorine	Mellon/Falconer	CG	19.12
1860	*Garibaldi, or, The Rival Patriots*	Frederic Cowen/Rosalind Cowen	Private Performance	?
	Romance	Leslie/Simpson	CG	02.02
	Lurline	Wallace/Fitzball	CG	23.02
	Robin Hood	Macfarren/ Oxenford	HM	11.10

YR	TITLE	COMPOSER/ LIBRETTIST	THEATRE	DATE
1860	*Bianca, the Bravo's Bride*	Balfe/Simpson	CG	06.12
	Queen Topaze	Linley/Oxenford	HM	26.12
1861	*The Amber Witch*	Wallace/Chorley	HM	28.02
	Ruy Blas	Glover/composer	CG	24.10
	Count and Cottage	Clay/Taylor	CG	11.11
	The Puritan's Daughter	Balfe/Bridgeman	CG	30.11
1862	*The Lily of Killarney*	Benedict/ Oxenford & Boucicault	CG	10.02
	Love's Triumph	Wallace/Planché	CG	03.11
	Law versus Love	Linley/composer	Prn	06.12
1863	*The Armourer of Nantes*	Balfe/Bridgeman	CG	12.02
	A Lord for an Hour	Saunders/ composer	TR Sheffield	17.07
	The Desert Flower	Wallace/Harris & Williams	CG	12.10
	Jessy Lea	Macfarren/ Oxenford	GI	02.11
	Blanche de Nevers	Balfe/Brougham	CG	21.11
1864	*She Stoops to Conquer*	Macfarren/ Fitzball	CG	11.02
	The Soldier's Legacy	Macfarren/ Oxenford	GI	10.07
	The Sleeping Queen	Balfe/Farnie	GI	31.08
	Helvellyn	Macfarren/ Oxenford	CG	03.11
	Rose, or, Love's Ransom	Hatton/Edwards	CG	26.11
	The Bride of Song	Benedict/Farnie	CG	03.12

Select Bibliography

PRIMARY SOURCES

Periodicals

Athenaeum
Gazetta Musicale di Milano
La France Musicale
Monthly Musical Record
Musical Times
Musical World
Punch, or the London Charivari
The Times

Books and Articles

Addison, Joseph, *The Spectator*, no.18 [21 March 1711]

Archer, William, *Masks or Faces? A Study in the Psychology of Acting* (London, 1888)

Arditi, Luigi, *My Reminiscences* (London, 1896)

Banister, Henry C., *George Alexander Macfarren: His Life, Works, and Influence* (London, 1891)

Barrett, William, *Balfe: his life and work* (London, 1882)

Beale, Willert, *The Light of Other Days*, 2 vols (London, 1890)

Benedict, Julius, *Weber* (London, 1881)

Berlioz, Hector, *Les Soirées de l'Orchestre* (Paris, 1852)

------*Oeuvres Littéraires Correspondance Générale*, edited by Pierre Citron, 5 vols (Paris, 1972-89)

------*The Memoirs of Hector Berlioz*, translated and edited by David Cairns (London, 1977)

Bernard, Bayle, *The Life of Samuel Lover R.H.A.*, 2 vols (London, 1874)

Bunn, Alfred, *The Stage: both before and behind the curtain*, 3 vols (London, 1840)
------*A Word with Punch* (London, [1847])
------*The Case of Bunn versus Lind ... February 22nd, 1848 ...* (London, 1848)
Cazalet, Rev. W.W., *The History of the Royal Academy of Music* (London, 1854)
Dallas, R.C., *Elements of Self-Knowledge* (London, 1802)
Darwin, Charles, *The Expression of the Emotions in Man and Animals* (London, 1872)
Davidson, Mrs Frances, 'Remarks,' in Eugene Scribe, *Giralda*, translated by Mrs Davidson (London, [1850]), pp.[5]-7
Day, Alfred, *Alfred Day's Treatise on Harmony, Edited, with an Appendix, by G.A.Macfarren* [2nd ed.] (London, 1885)
de Genlis, Stephanie Brulart, *La Siège de Rochelle*, 3 vols (Paris, 1808)
Dennis, John, *An Essay on the Opera's after the Italian Manner* (London, 1706)
Descartes, René, *Les Passions de l'Âme* (Paris, 1649)
Dickens, Charles, *Sketches by Boz*, Gadshill Edition, 2 vols (London, 1898)
------*The Letters of Charles Dickens*, The Pilgrim Edition (London, 1965--)
Diderot, Denis, *Essais sur la peinture* (Paris, 1795 [1765])
------and Jean Le Rond d'Alembert, *Encyclopédie ou dictionnaire raisonné des sciences, des arts et des métiers*, 17 vols (Paris, 1751-65)
Fitzball, Edward, *Thirty-Five Years of a Dramatic Author's Life*, 2 vols (London, 1859)
Griffin, Daniel, *Life of Gerald Griffin Esq. by his Brother, Daniel Griffin, M.D.* (London, 1843)
Harrison, Clifford, *Stray Records*, 2 vols (London, 1893)
Hazlitt, William, 'Sir Walter Scott, Racine, and Shakespear,' *The Complete Works*, edited by P.P.Howe, 21 vols (London & Toronto, 1930-34), 12:336-46
------'On the Opera,' *The Complete Works*, edited by P.P. Howe, 21 vols (London & Toronto, 1930-34), 20:92-96
Hogarth, George, *Memoirs of the Musical Drama*, 2 vols (London, 1838)
------*Memoirs of the Opera in Italy, France, Germany, and England*, 2 vols (London, 1851) [2nd ed. of *Memoirs of the Musical Drama*]

Holland, Henry Scott, and W.S. Rockstro, *Memoir of Madame Jenny Lind-Goldschmidt*, 2 vols (London, 1891)

Home, Henry, Lord Kames, *Elements of Criticism*, 3 vols (Edinburgh, 1762)

Hueffer, Francis, *Half a Century of Music in England* (London, 1889)

Hunt, Leigh, *Musical Evenings*, edited by David R.Cheney (Columbia, 1964)

Johnson, Dr Samuel, *Rambler*, Essay no.4, 31 March 1750

Kenney, Charles Lamb, *A Memoir of Michael William Balfe* (London, 1875)

Lavater, Johann Caspar, *Essays on Physiognomy*, translated by Thomas Holcroft, 18th ed. (London, [1885?])

Linley, George, *Musical Cynics of London: A Satire (Sketch the First)* (London, 1862)

------*The Modern Hudibras* (London, 1864)

Lockman, John, *Some Reflexions Concerning Operas, Lyric Poetry, Music, &c.* (London, 1740)

Logier, Johann, *System of the Science of Music, Harmony, and Practical Composition* (London, 1827)

Lumley, Benjamin, *Reminiscences of the Opera* (London, 1864)

Macfarren, George, 'Mendelssohn,' *Musical World*, 24 (1849): 33-36

------*Six Lectures on Harmony*, (London, 1867)

------'The Italian language: its evil influence upon music,' *Musical Times*, 14 (1869): 7-10

------*The Imperial Dictionary of Universal Biography*, edited by J.F.Waller, 3 vols (London, Glasgow, & Edinburgh, n.d.), s.v. 'Balfe, Michael William,' 1:357-58

Meyerbeer, Giacomo, *Briefwechsel und Tagebücher*, edited by Heinz Becker and Gudrun Becker, 4 vols (Berlin, 1960-85)

Osborne, G.A., 'Musical Coincidences and Reminiscences,' *Proceedings of the Musical Association*, 9 (1882-83): 95-113

Phillips, Henry, *Musical and Personal Recollections*, 2 vols (London, 1864)

Planché, James Robinson, *Recollections and Reflections*, 2 vols (London, 1872)

Pope, Alexander, *The Dunciad* (London, 1742)

Pougin, Arthur, *William-Vincent Wallace: étude biographique et critique* (Paris, 1866)

Pückler-Muskau, Hermann, *Tour in Germany, Holland and England in the Years 1826, 1827, & 1828*, [vols 3 and 4 of *Tour in England,*

Ireland, and France in the Years 1828 & 1829, 'by a German Prince'], 4 vols (London, 1832)
Radcliffe, Mrs Ann, *The Mysteries of Udolpho*, 4 vols (London, 1794)
Reeves, Sims, *My Jubilee, or, Fifty Years of Artistic Life* (London, [1889])
Report from The Select Committee appointed to inquire into the Laws affecting Dramatic Literature [London, 1832]
Rice, Charles, *The London Theatre in the Eighteen-Thirties*, edited by Arthur Colby Sprague and Bertram Shuttleworth (London, 1950)
Richardson, Samuel, *Pamela*, 2 vols (London, 1740)
Roche, Mrs Regina, *The Children of the Abbey*, 4 vols (London, 1796)
Salaman, Charles, 'On Musical Criticism,' *Proceedings of the Musical Association* (hereafter *PMA*), 2 (1875-76): 1-15
------'On Music as a Profession in England,' *PMA*, 6 (1879-80): 107-24
Scott, Sir Walter, 'Essay on the Drama,' in *Essays on Chivalry, Romance, and the Drama* (London, [1888]), pp.135-226
Shaw, Bernard, *Shaw's Music*, 2nd rev. edition by Dan. H. Laurence, 3 vols (London, 1989)
Squire, W. Barclay, *Dictionary of National Biography*, edited by Leslie Stephen, 68 vols (London, 1885-1900), s.v. 'Balfe, Michael William,' 3:44-47
St Leger, Heyward, *Reminiscences of Balfe* (London, [1871])
Twining, Thomas, *Aristotle's Treatise on Poetry, translated: with two Dissertations...* (London, 1789)
Wagner, Richard, *Wagner Writes from Paris*, edited and translated by Robert L.Jacobs and Geoffrey Skelton (London, 1973)

SECONDARY SOURCES

Abbiati, Franco, *Giuseppe Verdi*, 4 vols (Milan, 1959)
Adrian, Arthur A., *Mark Lemon: First Editor of Punch* (London, 1966)
Brooks, Peter, *The Melodramatic Imagination* (New York, 1985)
Burton, Nigel, 'Opera: 1865-1914,' in *The Romantic Age*, edited by Nicholas Temperley, volume 5 of *The Blackwell History of Music in Britain*, 6 volumes projected (Oxford, 1988), pp.330-57
Cambi, Luisa, 'Bellini. Un pacchetto di autografi,' in *Scritti in onore di Luigi Ronga* [edited by Riccardo Ricciardi], (Milan & Naples, 1973), pp.53-90

Carr, Bruce, 'The first all-sung English 19th-century opera,' *Musical Times*, 115 (1974): 125-26

------'Theatre Music: 1800-1834,' in *The Romantic Age*, edited by Nicholas Temperley, volume 5 of *The Blackwell History of Music in Britain*, 6 volumes projected (Oxford, 1988), pp.288-306

Fletcher, Ifan, 'Balfe and Barrymore at Milan,' *Theatre Notebook*, 7 (1953): 60-62

Forsyth, Cecil, *Music and Nationalism* (London, 1911)

Gatti, Carlo, *Il Teatro alla Scala*, 2 vols (Milan, 1964)

Hurd, Michael, 'Opera: 1835-1865,' in *The Romantic Age*, edited by Nicholas Temperley, volume 5 of *The Blackwell History of Music in Britain*, 6 volumes projected (Oxford, 1988), pp.307-29

Jacobs, Arthur, *Arthur Sullivan: A Victorian Musician* (2nd ed., Aldershot, 1992)

Langley, Leanne, 'The English Musical Journal in the Early Nineteenth Century' (unpublished Ph.D. dissertation, University of North Carolina at Chapel Hill, 1983)

------'The Life and Death of The *Harmonicon*: an Analysis,' *R[oyal] M[usical] A[ssociation] Research Chronicle*, 22 (1989): 137-63

Mackerness, E.D., 'Leigh Hunt's Musical Journalism,' *Monthly Musical Record*, 86 (1956): 212-22

Nicoll, Allardyce, *A History of English Drama 1660-1900*, 6 vols (Cambridge, 1952-59)

Rowell, George, *Queen Victoria Goes to the Theatre* (London, 1978)

Shattuck, Charles H., ed., *Bulwer and Macready* (Urbana, 1958)

Survey of London, general editor F.H.W. Sheppard, 41 vols (London, 1900-83), 35, *The Theatre Royal Drury Lane and The Royal Opera House Covent Garden*

Temperley, Nicholas, 'The English Romantic Opera,' *Victorian Studies*, 9 (1966): 293-301

------*New Grove Dictionary of Music*, edited by Stanley Sadie, 20 vols (London, 1980), s.v. 'Wallace, (William) Vincent,' 20:175-78

------'Raymond and Agnes,' *Musical Times*, 107 (1966): 307-10

------ed., *The Romantic Age 1800-1914*, volume 5 of *The Blackwell History of Music in Britain*, 6 volumes projected (Oxford, 1988).

------ed., *The Lost Chord: Essays on Victorian Music* (Bloomington and Indianapolis, 1989)

Trilling, Lionel, and Harold Bloom, eds., *Victorian Prose and Poetry* (New York, 1973)

Urwin, G.G., 'Bunn and his influence in the theatre' (unpublished

Ph.D. dissertation, University of London, 1956)
------'Alfred Bunn 1796-1860: A Revaluation,' *Theatre Notebook*, 11 (1957): 96-102

White, Eric Walter, *The Rise of English Opera* (London, 1951)

------*A History of English Opera* (London, 1983)

------*A Register of First Performances of English Operas and Semi-Operas* (London, 1983)

Whittall, Arnold, *Romantic Music* (London, 1987)

Index

This index lists references to persons and the titles of the operas of the main composers studied. Major references are noted in bold. Indirect references are noted in parentheses.